Bioeconomy and Global Inequalities

Maria Backhouse · Rosa Lehmann ·
Kristina Lorenzen · Malte Lühmann ·
Janina Puder · Fabricio Rodríguez ·
Anne Tittor
Editors

Bioeconomy and Global Inequalities

Socio-Ecological Perspectives on Biomass Sourcing and Production

SPONSORED BY THE

palgrave
macmillan

Editors
Maria Backhouse
Institute of Sociology, Friedrich Schiller
University Jena, Jena, Thüringen, Germany

Rosa Lehmann
Institute of Sociology, Friedrich Schiller
University Jena, Jena, Thüringen, Germany

Kristina Lorenzen
Institute of Sociology, Friedrich Schiller
University Jena, Jena, Thüringen, Germany

Malte Lühmann
Institute of Sociology, Friedrich Schiller
University Jena, Jena, Thüringen, Germany

Janina Puder
Institute of Sociology, Friedrich Schiller
University Jena, Jena, Thüringen, Germany

Fabricio Rodríguez
Institute of Sociology, Friedrich Schiller
University Jena, Jena, Thüringen, Germany

Anne Tittor
Institute of Sociology, Friedrich Schiller
University Jena, Jena, Thüringen, Germany

BMBF Junior Research Group "Bioeconomy and Inequalities—Transnational Entanglements and Interdependencies in the Bioenergy Sector" (Funding Code 031B0021). https://www.bioinequalities.uni-jena.de/en. The sole responsibility for this publication lies with the editors and authors.

ISBN 978-3-030-68946-9   ISBN 978-3-030-68944-5 (eBook)
https://doi.org/10.1007/978-3-030-68944-5

© The Editor(s) (if applicable) and The Author(s) 2021. This book is an open access publication.
Open Access This book is licensed under the terms of the Creative Commons Attribution 4.0 International License (http://creativecommons.org/licenses/by/4.0/), which permits use, sharing, adaptation, distribution and reproduction in any medium or format, as long as you give appropriate credit to the original author(s) and the source, provide a link to the Creative Commons license and indicate if changes were made.
The images or other third party material in this book are included in the book's Creative Commons license, unless indicated otherwise in a credit line to the material. If material is not included in the book's Creative Commons license and your intended use is not permitted by statutory regulation or exceeds the permitted use, you will need to obtain permission directly from the copyright holder.
The use of general descriptive names, registered names, trademarks, service marks, etc. in this publication does not imply, even in the absence of a specific statement, that such names are exempt from the relevant protective laws and regulations and therefore free for general use.
The publisher, the authors and the editors are safe to assume that the advice and information in this book are believed to be true and accurate at the date of publication. Neither the publisher nor the authors or the editors give a warranty, expressed or implied, with respect to the material contained herein or for any errors or omissions that may have been made. The publisher remains neutral with regard to jurisdictional claims in published maps and institutional affiliations.

Cover illustration: Maram_shutterstock.com

This Palgrave Macmillan imprint is published by the registered company Springer Nature Switzerland AG
The registered company address is: Gewerbestrasse 11, 6330 Cham, Switzerland

Acknowledgements

Scientific publications are always the result of collaborative work, discussions and mutual feedback. This edited volume entitled "Bioeconomy and Global Inequalities: Socio-Ecological Perspectives on Biomass Sourcing and Production" builds on an international workshop held between 25 and 27 June 2019 in Jena, Germany. The workshop was hosted by the Junior Research Group "Bioeconomy and Inequalities. Transnational Entanglements and Interdependencies in the Bioenergy Sector", which is funded by the German Federal Ministry of Education and Research (BMBF). Besides the editors of this volume, Larry Lohmann, Renata Motta, Siti Rahyla Rahmat, Hariati Sinaga, Tero Toivannen, and Virginia Toledo López enriched the workshop with paper presentations, which were partly elaborated as contributions for this edited volume. Yvonne Kunz, Éric Pineault, David Tyfield, and Thomas Vogelpohl, in their role as discussants, provided critical comments about the texts and contributed to in-depth discussions. Christin Bernhold, Emma Dowling, Samadhi Lipari and Oliver Pye facilitated the exchange of ideas. We are grateful for all their help and suggestions. Further, we would especially like to thank Ilka Scheibe, Philip Koch, Ronja Wacker and Louise Wagner for their administrative and organisational work before, during and after the workshop, as well as to all of the attendees who enriched several rounds of discussion with their critical and inspiring questions.

We are also grateful to Ronja Wacker, Maximilian Schneider and Laura Mohacsi for checking literature lists in this volume and adapting the manuscript to fulfil the publisher's guidelines, as well as to Simon Phillips for language editing and proofreading. We would like to send a special message of gratitude to Rachael Ballard and Joanna O'Neill from Palgrave Macmillan for their support and guidance throughout the publication process. Finally, we are very grateful to the BMBF for funding this volume as an open access publication: our book is about inequalities, and we consider it crucial that academic knowledge is accessible to everyone.

Jena
October 2020

Maria Backhouse
Rosa Lehmann
Kristina Lorenzen
Malte Lühmann
Janina Puder
Fabricio Rodríguez
Anne Tittor

Contents

Part I Introduction

1 Contextualizing the Bioeconomy in an Unequal World: Biomass Sourcing and Global Socio-Ecological Inequalities 3
Maria Backhouse, Rosa Lehmann, Kristina Lorenzen, Janina Puder, Fabricio Rodríguez, and Anne Tittor

Part II Rethinking the Bioeconomy, Energy, and Value Production

2 Global Inequalities and Extractive Knowledge Production in the Bioeconomy 25
Maria Backhouse

3 Neoliberal Bioeconomies? Co-constructing Markets and Natures 45
Kean Birch

4 Tools of Extraction or Means of Speculation? Making Sense of Patents in the Bioeconomy 65
Veit Braun

5 Bioenergy, Thermodynamics and Inequalities 85
Larry Lohmann

Part III Bioeconomy Policies and Agendas in Different Countries

6 Knowledge, Research, and Germany's Bioeconomy: Inclusion and Exclusion in Bioenergy Funding Policies 107
Rosa Lehmann

7 A Player Bigger Than Its Size: Finnish Bioeconomy and Forest Policy in the Era of Global Climate Politics 131
Tero Toivanen

8 Sugar-Cane Bioelectricity in Brazil: Reinforcing the Meta-Discourses of Bioeconomy and Energy Transition 151
Selena Herrera and John Wilkinson

Part IV Reconfigurations and Continuities of Social-ecological Inequalities in Rural Areas

9 *Buruh Siluman*: The Making and Maintaining of Cheap and Disciplined Labour on Oil Palm Plantations in Indonesia 175
Hariati Sinaga

10 Superexploitation in Bio-based Industries: The Case of Oil Palm and Labour Migration in Malaysia 195
Janina Puder

11 Sugarcane Industry Expansion and Changing Rural Labour Regimes in Mato Grosso do Sul (2000–2016) 217
Kristina Lorenzen

12 Territorial Changes Around Biodiesel: A Case Study
    of North-Western Argentina   239
    Virginia Toledo López

Part V The Extractive Side of the Global Biomass
        Sourcing

13 Contested Resources and South-South
    Inequalities: What Sino-Brazilian Trade Means
    for the "Low-Carbon" Bioeconomy   265
    Fabricio Rodríguez

14 Sustaining the European Bioeconomy: The Material
    Base and Extractive Relations of a Bio-Based
    EU-Economy   287
    Malte Lühmann

15 Towards an Extractivist Bioeconomy? The Risk
    of Deepening Agrarian Extractivism When
    Promoting Bioeconomy in Argentina   309
    Anne Tittor

Index   331

Notes on Contributors

Maria Backhouse is a Professor of Global Inequalities and Socio-ecological Change at the Institute of Sociology, Friedrich Schiller University Jena, Germany. She holds a Ph.D. in sociology and is Director of the Junior Research Group Bioeconomy and Inequalities funded by the German Federal Ministry of Education and Research (BMBF). Her current research focuses on unequal knowledge production in the global bioeconomy.

Kean Birch is an Associate Professor in the Faculty of Environmental and Urban Change at York University, Canada. His most recent books include *Assetization* (edited with Fabian Muniesa, MIT Press, 2020) and *Neoliberal Bio-economies?* (Palgrave Macmillan, 2019).

Veit Braun is a Research Associate at the Institute for Sociology, Goethe University Frankfurt. He studied sociology, political science, and environmental studies in Vienna and Munich. His Ph.D. (LMU Munich, 2018) focused on the changing role of property in plant breeding. He is currently part of the Cryosocieties ERC project in Frankfurt, where he investigates the frozen life of animal cells and DNA from conservation biobanks.

Selena Herrera is a Postdoctoral Researcher at the Research Unit Markets, Networks and Values, Rural Federal University of Rio de Janeiro, and a member of the Electricity Sector Research Group at the Institute of Economics, Federal University of Rio de Janeiro, Brazil. She holds a Ph.D. in energy planning and her current research focuses on the promotion of a sustainable transition in the Brazilian electricity sector.

Rosa Lehmann is a Postdoctoral Researcher at the Institute of Sociology, Friedrich Schiller University Jena, Germany. She holds a Ph.D. in political science and is a member of the Junior Research Group Bioeconomy and Inequalities, which is funded by the German Federal Ministry of Education and Research (BMBF). Her research covers conflict and inequalities related to renewable energies from a theoretical perspective based on political ecology.

Larry Lohmann has worked with social movements in Thailand, Ecuador and elsewhere. His books include *Pulping the South* (1996, with Ricardo Carrere), *Mercados de carbono: La neoliberalizacion del clima* (2012), and *Cadenas de bloques, automatizacion y trabajo* (2020). His articles have appeared in political economy, environment, geography, accounting, Asian studies, law, science studies, anthropology, development and socialist theory journals.

Kristina Lorenzen is a Researcher at the Institute of Sociology, Friedrich Schiller University Jena, Germany. She is a Latin Americanist (M.A.) and a member of the Junior Research Group Bioeconomy and Inequalities, which is funded by the German Federal Ministry of Education and Research (BMBF). Her current research focuses on labour and land relations in the Brazilian sugarcane sector.

Malte Lühmann is a Researcher at the Institute of Sociology, Friedrich Schiller University Jena, Germany. He is a political scientist and a member of the Junior Research Group Bioeconomy and Inequalities, which is funded by the German Federal Ministry of Education and Research (BMBF). Lühmann is specialized in European integration and global political economy; his recent research has been focused on transnational relations of the European bioeconomy and related political processes.

Janina Puder is a Researcher at the Institute of Sociology, Friedrich Schiller University Jena, Germany. She is a sociologist (M.A.) and a member of the Junior Research Group Bioeconomy and Inequalities, which is funded by the German Federal Ministry of Education and Research (BMBF). Her current research focuses on labour migration and overexploitation in the Malaysian palm oil sector.

Fabricio Rodríguez is a Postdoctoral Researcher at the Institute of Sociology, Friedrich Schiller University Jena, Germany. He holds a Ph.D. in political science and is a member of the Junior Research Group Bioeconomy and Inequalities, which is funded by the German Federal Ministry of Education and Research (BMBF). His current work focuses on resource trade, energy transitions, and the political economy of Chinese-Latin American relations.

Hariati Sinaga holds a Ph.D. from the University of Kassel, Germany. Her research interests include labour rights, labour relations and gender. Her current research focuses on labour relations in Indonesian oil palm plantations.

Anne Tittor is a Postdoctoral Researcher at the Institute of Sociology, Friedrich Schiller University Jena, Germany. She is a Sociologist, holds a Ph.D. in political science, and is a member of the Junior Research Group Bioeconomy and Inequalities, which is funded by the German Federal Ministry of Education and Research (BMBF). Her current research focuses on political ecology, socio-environmental conflicts and extractivism.

Tero Toivanen is a Postdoctoral Researcher at the BIOS research unit, Helsinki, Finland. He holds a Ph.D. in social and economic history. His research interests include the intertwined processes of capital, labour and ecology in concrete world-historical environments. Lately, his research has focused on the political economy of Finnish forestry, climate change, right-wing populism, and the governance of a low-carbon economic transition.

Virginia Toledo López is a Researcher at the National Scientific and Technical Research Council (CONICET). She holds a Ph.D. in social science, is based at the Institute of Studies of Social Development (INDES), and is also member of the Environmental Studies Group at the Gino Germani Institute (IIGG) of the Buenos Aires University (UBA), Argentina. Her research focuses on environmental conflicts regarding agribusiness expansion.

John Wilkinson is a Professor of Economic Sociology at the Graduate Center for Development, Agriculture and Society (DDAS/CPDA) and Director of the Research Unit Markets, Networks and Values at Federal Rural University of Rio de Janeiro, Brazil. He has published widely on food system issues and is currently researching the impacts of the new waves of food innovation and consumption practices.

List of Figures

Fig. 8.1   Total BNDES funds contracted by the biomass industry between 2007 and 2019, in million R$ (*Source* NovaCana [2020]. Authors' illustration)   163

Fig. 9.1   Employment structure on large-scale plantations (*Source* adapted from Siagian et al. 2011, p. 5)   183

Fig. 11.1   Employees in the sugarcane sector in Mato Grosso do Sul, 2007–2016 (*Source* RAIS, organised by DIESSE)   225

Fig. 11.2   Number of Employees in the sugarcane sector by area, Mato Grosso do Sul, 2007–2016 (Source RAIS, organised by DIESSE)   225

Fig. 12.1   Agrochemicals use (kg/lt) and soya Farmland (ha). 1990/1991–2016/2017 (*Source* Own elaboration, adapted from *Sistema de Datos Abiertos de la Secretaría de Agroindustria*, https://datos.agroindustria.gob.ar/dataset/estimaciones-agricolas and *Naturaleza de derechos* 2019)   243

Fig. 12.2	Biodiesel agroindustry in Argentina. Location in 2008 and 2012 (*Source* Own elaboration, adapted from *Secretaría de Energía*. See, https://www.argentina.gob.ar/produccion/energia/. Accessed 12 May 2015)	244
Fig. 12.3	Main destinations of Argentinian Biodiesel (t). Provisional data (*Source* Secretaría de Energía. See, https://www.argentina.gob.ar. Accessed 29 Oct 2019)	245
Fig. 13.1	Shifting dynamics in Global Energy Consumption, 1990–2018 [Mtoe] (*Source* Enerdata (2015, 2019); Global Energy Statistical Yearbook (2018), Accessed 1 April 2020. Author's illustration)	268
Fig. 13.2	Brazilian exports to China by commodity, 2000–2018 [Billion US$] (*Source* Chatham House (2018), 'resourcetrade.earth', http://resourcetrade.earth/. Accessed 1 April 2020. Author's illustration)	277
Fig. 14.1	Biomass inputs in the EU-28 over time (RMI in Million Tonnes of Raw Material Equivalent) (*Source* Eurostat)	294
Fig. 14.2	Biomass inputs in the EU-28 by type, 2016 (in Million Tonnes of Raw Material Equivalent) (*Source* Eurostat)	294
Fig. 14.3	Biomass imports to the EU-28 by country, 2016 (in Million Tonnes; Netweight for Wood) (*Source* UN Comtrade database; columns only show the biggest importers of each commodity with a combined proportion of at least 90% of imports for the respective commodity)	296

Part I

Introduction

1

Contextualizing the Bioeconomy in an Unequal World: Biomass Sourcing and Global Socio-Ecological Inequalities

Maria Backhouse, Rosa Lehmann, Kristina Lorenzen, Janina Puder, Fabricio Rodríguez, and Anne Tittor

The term 'bioeconomy' is commonly met with a sense of uncertainty regarding its meaning and purpose. In general, there are three different fields of public and scientific debate about the bioeconomy. Nicholas Georgescu-Roegen (1971) referred to the bioeconomy as a transformational pathway towards a degrowth society. In contrast, the debate about 'biocapitalism' focuses on the commodification of bodies, biological matters and micro-organisms in the context of biotechnological innovation (Cooper 2014; Sunder Rajan 2007). Lastly, bioeconomy policies are also viewed as presenting themselves as a means of replacing the fossil base of modern societies through the intensified use of biomass sources.

In this volume, we primarily refer to this third strand of the debate. Against the background of climate change, bioeconomy was introduced as a transitional strategy by the OECD in 2009 and was subsequently

---

M. Backhouse (✉) · R. Lehmann · K. Lorenzen · J. Puder · F. Rodríguez · A. Tittor
Institute of Sociology, Friedrich Schiller University, Jena, Germany
e-mail: maria.backhouse@uni-jena.de

revisited by Germany (BMBF and BMEL[1] 2020; BMBF 2010), the US (The White House 2012) and the EU (European Commission 2012, 2018). In these policy processes, the biotechnology sector has played (to varying degrees in different countries) an influential role in determining the content and direction of specific measures to facilitate the emergence and institutionalization of the bioeconomy (Meyer 2017). Many corresponding policy documents address primarily the agricultural and forest sectors while highlighting the significance of research and innovation (R&I) programmes as the pillars of a knowledge-based transition towards a sustainable bioeconomy. By 2018, 14 countries as well as the EU had adopted national bioeconomy strategies; another 34 countries refer to the bioeconomy in their agricultural or research strategies (German Bioeconomy Council 2018, p. 13).

Considering this landscape, the concept of the bioeconomy is far from being static or monolithic. There is no common definition of the bioeconomy, since the objectives of national or supranational policy strategies vary depending on the technical background and specializations of the actors involved, as well as on sector views and interests related to existing biomass and biotech industries (Kleinschmit et al. 2014; Backhouse et al. 2017; Vivien et al. 2019). In some cases, the prefix 'bio' stands for the promotion of biotechnologies (OECD 2009). In the case of the EU, it highlights the use of biomass as the resource base of a 'knowledge-based bioeconomy' (European Commission 2012), or a 'circular bioeconomy'[2] (id. 2018; BMBF and BMEL 2020). The strategies and policies of semi-peripheral countries such as Argentina or Malaysia can be placed between the biomass-focus of the EU and the biotech-focus of the OECD.

Despite their specificities, there is a common assumption and narrative enshrined in all of these strategies: the idea that technological innovations are a necessary means of decoupling[3] economic growth from the

---

[1] BMBF is the German abbreviation for Bundesministerium für Bildung und Forschung and means Federal Ministry of Education and Research. BMEL stands for Bundesministerium für Ernährung und Landwirtschaft or Federal Ministry of Food and Agriculture.

[2] According to the European Commission (2015), a circular economy refers to the use of and reuse of products, materials and resources for as long as possible as part of the economic circuit.

[3] On the impossibilities of a circular bioeconomy from a metabolic standpoint, see Giampietro (2019).

overexploitation of resources and the harmful levels of CO_2-emissions generated through capitalist modes of production, consumption and energy combustion.

Although bioeconomy policies address global problems, the political discussions and research on the emerging bioeconomy are mainly focused on Europe and North America (see Backhouse in this volume). This is particularly striking since the bioeconomy relies on growing levels of biomass production for food, fodder, fibres and bioenergy, as well as for chemical components for biotechnologies, which are produced worldwide. Yet, a global perspective that considers the production of globally traded biomass and its effects on the agricultural and forestry sectors of different countries as well as knowledge production in several contexts beyond Europe and North America is still a lacuna in the political and research fields on the bioeconomy.

With this edited volume, we seek to address this research gap insofar as we scrutinize bioeconomy policies in several countries in (and across) both the semi-peripheries and the centres. We consider interconnections between different world regions and assume that bioeconomy policies as well as their main fields of action (research and development, agriculture and forest sectors) are not developed and implemented within ahistorical vacuums. Instead, they are intertwined with global socio-ecological inequalities between centres and semi-/peripheral countries as well as within countries since colonial times. Hence, this volume seeks to contribute towards answering the following guiding questions: How is the bioeconomy dealt with in different countries? To what extent does the bioeconomy perpetuate or change existing global socio-ecological inequalities between biomass producing semi-peripheries and centres with regard to where processing takes place and value is produced?

We use the term *socio-ecological* to underline the assumption of political-economic approaches within the research field of political ecology that view nature and society as dialectically interrelated (Görg 2004). Nature cannot be thought of without society and vice versa. From this perspective, today's global socio-ecological inequalities are shaped by the capitalist mode of production: capitalism, with its need to accumulate and grow, has led to a level of resource depletion that is unparalleled in human history (O'Connor 1986), and it affects people and nature in unequal ways.

Drawing on theoretical and empirical research in political ecology, we identify four dimensions of global *socio-ecological inequalities*. (1) Resource access and use: people are not only unequally integrated as paid or non-paid labour into the production and reproduction processes of global capitalism, but they are also asymmetrically involved in the (over)use of natural resources. As research on unequal ecological exchange and unequal ecological footprints show, this socio-ecological inequality has a global dimension, since resource use and consumption by individuals is influenced by their place of residence as well as whether they live in semi-peripheries or capitalist centres (Bunker 1985; Martinez-Alier et al. 2016). (2) Environmental degradation: as environmental and climate justice movements as well as ecofeminists demonstrate at the local to the global level, people are also unequally exposed to the negative consequences of the degradation of nature, such as damage to health by pesticides. Further, these inequalities are re/produced along different structural categories such as class, gender, ethnicity and/or citizenship that influence and reinforce each other (Agarwal 1998; Bullard 2000; Acselrad 2010; Sundberg 2008). (3) Unequal production of knowledge: studies on green growth policies such as the promotion of renewables, or on conservation projects show that people are unequally involved in the political processes of problem definition and developing technical solutions (e.g. Escobar 1998; Lehmann 2019). As a result, (4) the changes that this leads to, such as the expansion of palm oil plantations for biodiesel, often have negative impacts on marginalized classes and groups such as small farmers or indigenous peoples as they usually lack the means to defend their land and customary rights (e.g. Backhouse 2016; Fairhead et al. 2012; Tittor 2020).

The *global perspective* is of utmost importance, since the globalized agricultural and forest sectors are inserted directly and indirectly into the unequal global relations that have evolved since colonial times (Bunker 1985; Moore 2000). We draw on the insights of world systems theory that social inequalities cannot solely be explained on a national level since they are shaped also by inequalities between countries (Korzeniewicz and

Moran 2012). In this perspective, global inequalities need to be analysed on a global scale that includes a historical perspective of 500 years of capitalism, and an understanding that colonialism enabled capitalism and structured global uneven developments (Wallerstein 2007). In this light, we address the pitfalls of methodological nationalism. While the nation state remains important in the introduction, construction, socialization, implementation, maintenance, legitimation and even defence of many bioeconomy agendas, the study of how biomass, and particularly bioenergy, is to offset societal change in times of global ecological crises requires an analytical move that goes beyond the study of national 'containers'.

Against this background, we have divided the two guiding questions into four blocks. Each chapter in this volume addresses at least one of the following questions:

- How can we think and/or rethink the concepts of bioeconomy and energy? How can a global perspective on socio-ecological inequalities contribute to a complex and critical understanding of bioeconomy?
- How is the bioeconomy discussed and implemented in different countries? Who participates in the negotiation of specific bioeconomy policies and who does not? Who determines the agenda?
- To what extent does the bioeconomy and biomass sourcing change or reproduce existing socio-ecological inequalities in rural areas?
- What are the implications of bioeconomy policies and transitions for existing relations of extraction and inequalities across regions?

The empirical focus of the volume mainly addresses the use of biomass and bioenergy by drawing on different analytical perspectives about the agricultural and forestry sectors. We refer to bioenergy as the use of biomass for producing fuels, i.e. first- and second-generation agrofuels, power and heat. Biomass-driven energy development in the transport, electricity and heating/cooling sector provides a large and longstanding depository of experiences that can be used to mobilize knowledge for the analysis of the bioeconomy. Energy is one of the pillars of many bioeconomy strategies. At the same time, bioenergy has been one of the focal points for social struggles surrounding the transition away from fossil-based resources. Thus, experiences in this field shed light on the transformation towards a post-fossil society, its actor constellations, challenges and contradictions.

The regional focus of this volume brings together multi-disciplinary contributions from social scientists working on bioeconomy-related issues in South and North America, East and Southeast Asia, and Europe. The volume has been organized by the German research group *Bioeconomy and Inequalities. Transnational Entanglements and Interdependencies in the Bioenergy Sector* (*BioInequalities*), which is funded by the BMBF. Therefore, it is worth noting that half of the authors are located in Germany. We acknowledge our positionalities within the academic structures of the Global North, an area of the world that is highly involved in the promotion of the international bioeconomy debate and its agenda. This volume is further enriched by a series of contributions from authors from and/or based in a variety of other countries and regions. The aim was to broaden the largely Eurocentric research landscape and political debate on bioeconomy, while moving discussions beyond the study of Europe and North America. We have included regions and countries that qualify as initiators of the bioeconomy debate and that have bioeconomy policies which are being put into practice. Due to the focus on bioenergy and biomass, longstanding important players in the transnational bioenergy sector have also been selected. We consider this volume a first impulse to expand the debate on bioeconomy, especially in terms of the impacts and forms of biomass and bioenergy production, and to encourage a regionally more varied research agenda that will hopefully include countries and regions that we could not consider here.

In this volume, we study global socio-ecological inequalities on various scales and consider different analytical categories. This multidimensionality of inequalities requires different methodological approaches. Thus, the contributions in this volume embrace a variety of methods: most chapters are based on qualitative research, including fieldwork, expert interviews and participatory observation. Many contributions complement their analyses with existing quantitative data sets. Some of the chapters analyse policy papers, expert and media debates on bioeconomy and bioenergy, while others refer more to socio-ecological change and the way it affects different social groups. Others put more emphasis on the historical emergence of inequalities and/or engage with ongoing debates about sustainability, energy, neoliberal natures, intersectionality and extractivism.

In the following, we outline the chapters' responses to the four blocks of questions of this volume. First, we approach central issues of the bioeconomy from different directions, such as unequal knowledge production, its neoliberal orientation, the production of value and unquestioned assumptions about (bio-)energies. In Sect. 1.2, we sketch out the main findings on bioeconomy policies in different countries. In Sect. 1.3, we summarize the reconfigurations and continuities of the socio-ecological inequalities that are present on the ground. In Sect. 1.4, we look at the extractive side of global biomass sourcing. Finally, we discuss the need for further research and the political implications of this volume.

1.1 Rethinking the Bioeconomy, Energy, and Value Production

The national and supranational strategies that target bioeconomy are mainly research funding strategies. The explicit aim of most national strategies is to compete for technological leadership in the emerging global knowledge-based bioeconomy. While many researchers criticize the technocratic and ecological modernization approach of bioeconomy policies, the global dimension of competing and unequal knowledge production beyond Europe and North America is still a research gap. As Maria Backhouse argues, the strategy-papers of the EU, Germany and the OECD reproduce global unequal knowledge production and simultaneously strengthen 'extractive knowledge' in the globalized agribusiness sector. Therefore, the bioeconomy concept is more connectable to the Brazilian agribusiness sector and less to agroecological movements and, thus, threatens to reproduce regional and global socio-ecological inequalities and aggravate climate change.

Kean Birch focuses on the market-based approach to the bioeconomy. He differs with critical perspectives that speak much too precipitately of a 'neoliberalization of nature', leaving little room to develop alternative bioeconomy approaches. Instead, he proposes to examine in detail how markets and nature are co-constructed, in other words, how the biophysical materialities of biomass intertwine with socio-economic configurations to produce different kinds of bioeconomies.

Another key question in the critical debate on the bioeconomy is about the extent to which the bioeconomy opens new ways of value creation and the role that patents play in this (Birch and Tyfield 2013): Are patents tools of extraction or speculation? Referring to the European vegetable market, Veit Braun's answer is that neither description applies completely to these patents. Braun argues that native trait patents are a legacy of biotech plant patents from the 1980s and 1990s, but that they follow different material, legal and economic logics. Thus, unlike GMO patents, native trait patents cannot be understood as tools for extracting surplus value from farmers. Instead, they are simply a means to capture investment on the stock markets. Braun concludes that there is no single business model that would explain the rush of companies to patent in conventional plant breeding. Therefore, patents must be understood as complex value objects that fulfil different functions for different actors, and that often defy their original purpose of stimulating and protecting innovation.

Larry Lohmann takes a step back from the guiding questions of this volume and radically criticizes the concept of energy that has become generally accepted in everyday vocabulary—even by critical scientists and social movements. He argues that any serious study of bioenergy and global inequalities must take account of the oppression inherent in thermodynamic energy itself. Thus, he first underlines that the abstract nature that we now call energy was organized during the nineteenth century in conjunction with new waves of capitalist mechanization centred on labour control and productivity. He then sketches some of the ways in which the social or ecological contradictions of thermodynamic energy are intensified in the twenty-first-century bioeconomy, suggesting that this is a useful framework for understanding many of the conflicts explored elsewhere in this book. Finally, the chapter draws out some of the implications for social movements and how they might place themselves more strategically in struggles over today's bioeconomy.

1.2 Bioeconomy Policies and Agendas in Different Countries

In our own studies of the bioeconomy in Germany, Malaysia, Brazil and Argentina, we noticed that few people outside of state expert circles can make sense of the term bioeconomy. Accordingly, we asked ourselves whether the bioeconomy is fact or fiction. We learned that bioeconomy agendas have been materializing in research funding policies and state incentives for bioenergy policies in all of the countries under study in the last ten years. However, these policies have been developed in expert fora and are mainly defined by dominant agribusiness, biotechnology and conventional forest sectors. The dominance of these sectors stands out in all the cases we present in this volume, from Finland, Brazil, Argentina, Malaysia and Indonesia to Germany and the European Union. Most contributions observe more or less cooperative relationships between state institutions and business associations, and a deliberate interest in expanding the production and commercialization of biomass products, biotechnologies and bioenergy sources. In Brazil, Indonesia and Malaysia, the bioeconomy has been appropriated by agribusiness sectors. This is exemplified by Anne Tittor in her analysis concerning Argentina. Tittor argues that the bioeconomy narrative has been appropriated by the agribusiness and biotechnology sectors, and that they use it to reframe their activities as sustainable. These actors are responsible for focusing the country's entire economy on soybean exports, while ignoring the negative social and environmental impacts.

Non-industrial actors focusing on small-scale agriculture, forestry management, or cooperative bioenergy production are absent in most policy processes (see Lehmann in this volume). Moreover, little to no concern is expressed about the integration or even protection of localized livelihoods (see Toledo López in this volume), where work and land issues (see Lorenzen, and Puder in this volume) as well as gender relations (see Sinaga in this volume) are directly—and negatively—affected by biomass sourcing. Thus, these contributions suggest that current bioeconomy policies do not provide sufficient entry points to enable alternative designs to become part of the process. This is partly due to the fact that the policy development process is not the subject of social debates about the form and objectives of this global socio-ecological transition project. This confirms on a global level what other authors have already discussed in the European context (TNI and Hands on the Land 2015): the bioeconomy is an exclusive project that lacks a democratic mechanism to ensure an open-ended negotiation process and the participation of all stakeholders.

For the German context, Rosa Lehmann emphasizes that the national bioeconomy agenda has thus far failed to integrate and reinvigorate the pre-existing knowledge and practices of civil-society actors engaged in cooperative schemes promoting citizen-based bioenergy production. Addressing issues of knowledge production from an energy justice perspective, Lehmann argues that the inclusion of these experiences would be a fundamental step towards the construction of a bioeconomy agenda that not only aims to induce technological change, but also to stimulate societal change.

Nevertheless, bioeconomy policy processes are contested and dynamic—and therefore changeable (Böcher et al. 2020). In this sense, the enduring intervention of civil society and critical academics have, for instance, led to some shifts in the revised version of the German bioeconomy strategy paper (BMBF and BMEL 2020). The paper acknowledges the fact that the additional need for biomass could aggravate the global socio-ecological crisis. Further, it opens its research funding explicitly to research and development in agroecology (ibid.). Whereas in the past, many official bioeconomy publications were full of euphonic promises of bioeconomy bringing sustainability and jobs, and mitigating climate change, a recent monitoring report questions Germany's growing ecological footprint, particularly if the country is to implement its new bioeconomy policy (Bringezu et al. 2020).[4]

---

[4] The monitoring report shows that the German economy is systematically based on the import of biomass and thus on the import of agricultural land and water: 16.7 million hectares (ha) are used within Germany, whereas abroad, Germany uses about 43 million ha of land. A substantial amount of the biomass produced for Germany comes from Asia, Africa and South and Central America—together, this is more than the amount produced by Germany and Europe itself (Bringezu et al. 2020, p. 87). The climate footprint of the agricultural goods consumed in Germany also exceeds total territorial emissions, which means that the emissions occur in the countries where the goods are produced. According to the monitoring projections, this climate footprint will hardly change until 2030.

In his study of Finland, Tero Toivanen shows that the bioeconomy can also become the subject of public controversy. In Finland, the bioeconomy has been adopted by the forestry sector. The dominant narrative paints the Finish forestry sector as sustainable, and as offering the country an important role within a European bioeconomy future. However, scientists and climate activists have challenged this view by arguing that increased forest harvesting will undermine Finland's climate objectives. In doing so, they have triggered a contentious public debate about the pros and cons of the bioeconomy.

In their sectoral analysis on sugarcane electricity in Brazil, Selena Herrera and John Wilkinson show that the promises made about the merits of second-generation biofuels and electricity produced from residues are far from materializing. Although sugarcane bioelectricity is framed as contributing to the diversification and distribution of power generation in Brazil, its development depends on specific public policies, and it faces hard competition from both the powerful fossil oil and gas sectors, and the renewable energy sector, which includes both solar and wind sources.

1.3 Reconfigurations and Continuities of Socio-Ecological Inequalities in Rural Areas

As various chapters in this volume outline, bioeconomy policies reproduce or reconfigure socio-ecological inequalities in the agricultural sector. The dominance of agribusiness in the development and implementation of most policy strategies and the absence of other stakeholders with alternative visions risks perpetuating existing socio-ecological inequalities in the agricultural sector in different countries as various qualitative studies in this volume demonstrate.

Kristina Lorenzen studies the changes to rural land and labour relations associated with sugarcane industry expansion in the Brazilian state of Mato Grosso do Sul. The expansion was encouraged by national policies that reflected global green development narratives. In this context, Brazilian sugarcane-based bioethanol was framed and reoriented as a climate-friendly alternative to fossil fuels. Nevertheless, this 'green industry' resulted in the reconfiguration of rural social inequalities. Sugarcane expansion contributed to the deceleration of agrarian reform, increased the integration of (non-indigenous) peasants as temporary wage workers, and led to a double exclusion of indigenous people from land and wage labour.

Similar dynamics that reinforce existing positions of social disadvantage in the production of biomass can be witnessed in the case of the steadily growing palm oil sector in Southeast Asia. Indonesia and Malaysia are by far the largest palm oil producers worldwide. Despite claims by both countries that palm oil production can be environmentally sustainable and, therefore, contribute significantly to climate protection and stop ecological degradation, and improve people's working and living conditions in the region, the evidence suggests otherwise. In her chapter, Janina Puder argues that migrant workers deployed to perform the physically most demanding and worst paid jobs in the industry are systematically overexploited to keep palm oil highly profitable for Malaysian producers. Puder's main argument is that the specific intersection of class and citizenship enables the overexploitation of migrant workers, and that this shows that bioeconomy developments do not necessarily break with key features of capitalism. A related argument is made by Hariati Sinaga. In her historically informed study of gendered labour in the Indonesian palm oil industry, Sinaga demonstrates that the customary forms of female labour on the plantations today evolved from the colonial period and continue to shape a cheap and disciplined female labour subject.

By examining the biodiesel sector in Argentina, Virginia Toledo López addresses the territorial impacts of biomass production in the Argentinian north. Toledo López' contribution puts the contradictions of Argentinian agrofuel production at centre stage. On the one hand, she identifies a strong developmentalist narrative related to bioenergy production; on the other, she argues that the production regions are confronted with the negative impacts of biomass production, whereas the products are sold on the world market. This connects the northern Argentinian peripheries to the centres. Thereby Toledo López shows that territorial inequalities are part and parcel of a bioeconomy situated in unequal structures.

1.4 The Extractive Side of Global Biomass Sourcing

It is still too early to say how the bioeconomy will affect the unequal global relations between the centres and semi-/peripheries in the long term. However, most policy papers are not aimed at changing the inequalities in global knowledge production or the global division of labour, and, instead, merely reproduce the status quo (see Backhouse in this volume). Further, Bringezu et al. (2020), who modelled the impact of the bioeconomy on biomass sourcing, suggest that the additional demand for biomass will amplify asymmetries between producing and processing countries. Therefore, the question is how these global socio-ecological inequalities will be changed by the rise of the BRICS-states.[5] For several decades now, the emergence of new global players including the BRICS, has been challenging the long-lasting dominance of countries that have represented the centre, both in political and economic terms. China and Brazil are significant examples of this shift. As Fabricio Rodríguez discusses in his chapter, the rise of these new heavyweights has had a significant impact on the direction of the global bioeconomy and, therefore, on the emergence of new global South-South inequalities. As Brazil intends to become an important supplier of bio-based resources and technologies, China's current role as a major consumer of non-renewable energies has created important constraints on the development of a global bioeconomy, while paving the way for new socio-ecological inequalities surrounding resource extraction.

However, the shifts in the global power structure do not mean that the global inequalities between the old centres and semi-/peripheries have become obsolete. For example, if we take a closer look at the quantity of resources that the EU will need in transitioning towards a bioeconomy, it is obvious that existing asymmetries in political and economic

---

[5] Brazil, Russia, India, China and South Africa.

power within the world system are not being called into question. Malte Lühmann contends that the often-criticized focus on resource extraction as a development strategy is likely to be reinforced by the material base of the EU bioeconomy scenarios. Furthermore, Lühmann argues that the EU's move towards a bioeconomy represents a continuation of the extractive relations that already shape the global market and the structural asymmetries between semi-/peripheral regions and industrial centres. Whereas European countries concentrate on the development of bio-based technologies and innovation, countries in economically weaker positions are tempted to compete to become important resource suppliers.

Such a dynamic can be witnessed, for example, in the case of Argentina, which is one of the world's largest producers and exporters of soy. The Argentinian government aims to place the country in an economically more favourable position on the global market and hopes to become a regional forerunner in terms of promoting biotechnologies. Furthermore, in addition to their exclusive character, bioeconomy policies also reproduce the existing agricultural model. In this sense, Anne Tittor refers to the Argentinian approach to bioeconomy as an 'extractivist bioeconomy'. In doing so, Tittor shows how concerns of ecological sustainability are sacrificed in the global race for profit and pioneer status in the global bioeconomy.

1.5 Outlook

In the light of current dynamics in different arenas of socio-ecological inequalities, the insights gained from European, South American and Southeast Asian cases underline that the bioeconomy, in its current form, is likely to reinforce or even produce new socio-ecological inequalities. Our findings have led us to identify four areas of further research.

First, more countries should be part of research agendas on the bioeconomy and bioeconomy-related issues. We believe that research should not only focus on countries that embrace some sort of explicit bioeconomy policy (programmes, laws, agendas), but also on evaluating the possibility of renewing biomass-centred policies in countries with a history in the relevant sectors. Examples include forest and agricultural resources in Russia and the former Soviet bloc, and countries in Sub-Saharan Africa.

Second, in our view, research should go beyond the analysis of strategy papers, narratives and the euphonic promises of the bioeconomy as sustainable, creating jobs and mitigating climate change and towards the analysis of different spheres of socio-ecological inequalities. For instance, these could include unequal access to and control over land or participation in policy development processes. As the chapters by Toledo López and Sinaga show, contextualizing the concrete socio-environmental impact of biomass sourcing both within specific local circumstances and historical structures reveal how this contributes to issues such as the devaluation and destruction of peasant livelihoods and makes female labour invisible. With contributions in this volume engaging with world systems theory, extractivism and research on transnational labour migration, we suggest how research could conceptualize the global perspective on socio-ecological inequalities from the local to the global level. However, more conceptual and empirical research is needed on the global dimension of socio-ecological inequalities in order to gain a deeper understanding of the interdependencies and interconnections between different societies, classes and groups that goes beyond the nation state. Insights from post- and decolonial research and border thinking provide starting points for further studies about a transnational bioeconomy, highlighting the coloniality/modernity of such an approach and the different axes of social inequalities.

Third, the real existing bioeconomy is currently strengthening powerful actors and mainstream practices in the forest and agricultural sector and, therefore, can contribute to deepening relations of exploitation, marginalization and dispossession as well as extractive and unequal trade relations. Against this background, we see a strong need to develop a transformative vision of the bioeconomy. One starting point could be the discussion about the meaning of bioeconomy as coined by Georgescu-Roegen as a radical degrowth perspective. Furthermore, we suggest conducting more research into existing alternative knowledge and practices. If Birch in this volume claims that another bioeconomy is possible, then the practices on which such a bioeconomy could rely on need to be examined in more detail—and this should include actors working on alternative innovations and through cooperative practices under a market-logic as much as actors who consider themselves to be working in line with and those who go beyond a growth imperative. If Lohmann challenges the common understanding of energy and the role of energy science in capitalism with little-e-energies, then a closer look should examine these little-e-energies—the actors and practices—in order to understand potential starting points for a societal transformation towards a more just and low-carbon bioeconomy.

Fourth, we need a broader public debate and negotiation about the objectives of the bioeconomy. Different authors in this volume call for the conceptual and political integration of civil-society based experiences and knowledges. This means including various actors in bioeconomy policy making as well as research funding, regardless of whether they are working explicitly towards an alternative bioeconomy. The environmental, climate and energy justice movements, peasant organizations like La Via Campesina at the global level, prosumer cooperatives, environmental, feminist and antiracist activists, workers in environmentally harmful as well as biomass-based industries, unions as well as 4future-groups on the regional and local level are just some examples. It is both scientifically and politically important to consider these actors as key players of the politics of the bioeconomy.

References

Acselrad, H. (2010). Ambientalização das lutas sociais – o caso do movimento por justiça ambiental. *Estudos Avançados, 24*(68), 103–119.

Agarwal, B. (1998). The Gender and Environment Debate. In R. Keil, D.V.J. Bell, P. Penz, & L. Fawcett (Eds.), *Political Ecology. Global and Local* (pp. 193–219, Innis centenary series). London, New York: Routledge.

Backhouse, M. (2016). The Discursive Dimension of Green Grabbing: Palm Oil Plantations as Climate Protection Strategy in Brazil. *Pléyade. Revista de Humanidades y ciencias Sociales, 18*, 131–157.

Backhouse, M., Lorenzen, K., Lühmann, M., Puder, J., Rodríguez, F., & Tittor, A. (2017). Bioökonomie-Strategien im Vergleich. Gemeinsamkeiten, Widersprüche und Leerstellen. Working Paper 1, Bioeconomy & Inequalities, Jena. https://www.bioinequalities.uni-jena.de/sozbemedia/neu/2017-09-28+workingpaper+1.pdf. Accessed 17 May 2020.

Birch, K., & Tyfield, D. (2013). Theorizing the Bioeconomy: Biovalue, Biocapital, Bioeconomics or … What? *Science, Technology, & Human Values, 38*(3), 299–327.

BMBF (2010). *National Research Strategy BioEconomy 2030. Our Route Towards a Biobased Economy.* Bonn. https://www.pflanzenforschung.de/application/files/4415/7355/9025/German_bioeconomy_Strategy_2030.pdf. Accessed 8 Aug 2020.

BMBF & BMEL (2020). *National Bioeconomy Strategy.* Berlin. https://biooekonomie.de/en/service/publications. Accessed 30 Oct 2020.

Böcher, M., Töller, A.E., Perbandt, D., Beer, K., & Vogelpohl, T. (2020). Research Trends: Bioeconomy Politics and Governance. *Forest Policy and Economics, 118*, 102219.

Bringezu, S., Banse, M., Ahmann, L., Bezama, A., Billig, E., Bischof, R., et al. (2020). Pilotbericht zum Monitoring der deutschen Bioökonomie. Kassel University, Center for Environmental Systems Research (CESR). https://kobra.uni-kassel.de/handle/123456789/11591. Accessed 8 Aug 2020.

Bullard, R. (2000). *Dumping in Dixie. Race, Class, and Environmental Quality* (3rd ed.). Boulder, Oxford: Westview.

Bunker, S.G. (1985). *Underdeveloping the Amazon: Extraction, Unequal Exchange, and the Failure of the Modern State.* Chicago, London: University of Chicago Press.

Cooper, M. (2014). Leben jenseits der Grenzen. Die Erfindung der Bioökonomie. In A. Folkers & T. Lemke (Eds.), *Biopolitik: Ein Reader* (pp. 468–524, Suhrkamp Taschenbuch Wissenschaft, Vol. 2080). Berlin: Suhrkamp.

Escobar, A. (1998). Whose Knowledge, Whose Nature? Biodiversity, Conservation, and the Political Ecology of Social Movements. *Journal of Political Ecology, 5*, 53–82.

European Commission (2012). *Innovating for Sustainable Growth: A Bioeconomy for Europe.* Luxembourg: Publications Office of the European Union. https://op.europa.eu/en/publication-detail/-/publication/1f0d8515-8dc0-4435-ba53-9570e47dbd51.%20Accessed%206%20Nov%202020. Accessed 6 November 2020.

European Commission (2015). *Communication from the Commission to the European Parliament, the Council, the European Economic and Social Committee and the Committee of the Regions: Closing the Loop - An EU Action Plan for the Circular Economy.* Brussels. https://eur-lex.europa.eu/resource.html?uri=cellar:8a8ef5e8-99a0-11e5-b3b7-01aa75ed71a1.0012.02/DOC_1&format=PDF. Accessed 20 Aug 2020.

European Commission (2018). *A Sustainable Bioeconomy for Europe: Strengthening the Connection Between Economy, Society and the Environment.* Brussels. https://ec.europa.eu/transparency/regdoc/rep/1/2018/EN/COM-2018-673-F1-EN-MAIN-PART-1.PDF. Accessed 17 May 2020.

Fairhead, J., Leach, M., & Scoones, I. (2012). Green Grabbing: A New Appropriation of Nature? *Journal of Peasant Studies, 39*(2), 237–261.

Georgescu-Roegen, N. (1971). *The Entropy Law and the Economic Process.* Harvard: Harvard University Press.

German Bioeconomy Council (2018). *Bioeconomy Policy (Part III). Update Report of National Strategies around the World.* Berlin. http://biooekonomierat.de/fileadmin/Publikationen/berichte/GBS_2018_Bioeconomy-Strategies-around-the_World_Part-III.pdf. Accessed 4 March 2019.

Giampietro, M. (2019). On the Circular Bioeconomy and Decoupling: Implications for Sustainable Growth. *Ecological Economics, 162,* 143–156.

Görg, C. (2004). The Construction of Societal Relationships with Nature. *Poiesis & Praxis, 3*(1), 22–36.

Kleinschmit, D., Lindstad, B.H., Thorsen, B.J., Toppinen, A., Roos, A., & Baardsen, S. (2014). Shades of Green: A Social Scientific View on Bioeconomy in the Forest Sector. *Scandinavian Journal of Forest Research, 29,* 402–410.

Korzeniewicz, R.P., & Moran, T.P. (2012). *Unveiling Inequality: A World-Historical Perspective.* London: Russell Sage Foundation.

Lehmann, R. (2019). *Der Konflikt um Windenergie in Mexiko.* Wiesbaden: Springer VS.

Martinez-Alier, J., Temper, L., Del Bene, D., & Scheidel, A. (2016). Is There a Global Environmental Justice Movement? *The Journal of Peasant Studies, 43,* 731–755.

Meyer, R. (2017). Bioeconomy Strategies: Contexts, Visions, Guiding Implementation Principles and Resulting Debates. *Sustainability, 9,* 1031.

Moore, J.W. (2000). Sugar and the Expansion of the Early Modern World-Economy: Commodity Frontiers, Ecological Transformation, and Industrialization. *Review (Fernand Braudel Center), 23*(3), 409–433.

O'Connor, J. (1986). Capitalism, Nature, Socialism: A Theoretical Introduction. *Capitalism, Nature, Socialism*, *1*(1), 11–38.

OECD (2009). *The Bioeconomy to 2030. Designing a Policy Agenda.* https://read.oecd-ilibrary.org/economics/the-bioeconomy-to-2030_9789264056886-en. Accessed 6 Nov 2020.

Sundberg, J. (2008). Tracing Race: Mapping Environmental Formations in Environmental Justice Research in Latin America. In D.V. Carruthers (Ed.), *Environmental Justice in Latin America* (pp. 24–47). Cambridge, London: The MIT Press.

Sunder Rajan, K. (2007). *Biocapital: The Constitution of Postgenomic Life* (2nd ed.) Durham: Duke University Press.

The White House (2012). *National Bioeconomy Blueprint.* Washington, DC. https://obamawhitehouse.archives.gov/sites/default/files/microsites/ostp/national_bioeconomy_blueprint_april_2012.pdf. Accessed 11 Feb 2020.

Tittor, A. (2020). Land. In O. Kaltmeier, A. Tittor & D. Hawkins (Eds.), *The Routledge Handbook to the Political Economy and Governance of the Americas* (pp. 159–172, Routledge Handbooks). New York, London: Routledge.

TNI and Hands on the Land (2015). The Bioeconomy. A Primer. https://www.tni.org/files/publication-downloads/tni_primer_the_bioeconomy.pdf. Accessed 1 Sep 2018.

Vivien, F.-D., Nieddu, M., Befort, N., Debref, R., & Giampietro, M. (2019). The Hijacking of the Bioeconomy. *Ecological Economics*, *159*, 189–197.

Wallerstein, I. (2007). *World-Systems Analysis: An Introduction* (5th ed., A John Hope Franklin Center Book). Durham: Duke University Press.

Open Access This chapter is licensed under the terms of the Creative Commons Attribution 4.0 International License (http://creativecommons.org/licenses/by/4.0/), which permits use, sharing, adaptation, distribution and reproduction in any medium or format, as long as you give appropriate credit to the original author(s) and the source, provide a link to the Creative Commons license and indicate if changes were made.

The images or other third party material in this chapter are included in the chapter's Creative Commons license, unless indicated otherwise in a credit line to the material. If material is not included in the chapter's Creative Commons license and your intended use is not permitted by statutory regulation or exceeds the permitted use, you will need to obtain permission directly from the copyright holder.

Part II

Rethinking the Bioeconomy, Energy, and Value Production

2

Global Inequalities and Extractive Knowledge Production in the Bioeconomy

Maria Backhouse

2.1 Introduction

National and supra-national bioeconomy strategies are primarily geared towards research promotion and funding. The common goal is to create a new, knowledge-based growth market that will mitigate the ecological crises, such as climate change, through technological means (Backhouse et al. 2017). Technological innovation, in this perspective, is the engine of the aspired transition from a fossil-based society to one based on biomass. Depending on the respective supra-national or national context, different technological domains take centre stage in strategy papers—from biotechnologies to the efficient cascading use of renewable biomass (ibid.; Kleinschmit et al. 2014). All bioeconomy strategies essentially rest on the belief that economic growth can be decoupled from excessive resource depletion through the production of knowledge and (bio-)technological innovation.

M. Backhouse (✉)
Institute of Sociology, Friedrich Schiller University, Jena, Germany
e-mail: maria.backhouse@uni-jena.de

This technology optimism, however, is problematic, as it fails to recognise that knowledge and technologies are by no means ahistorical, neutral or objective, but instead are socio-historically and locally embedded (Haraway 1988). Moreover, the scientific descriptions of the ecological crisis and the political and technology-based strategies to solve it are permeated by global relations of power and social inequalities that have evolved alongside the emergence of colonialism and capitalism. In this chapter, I seek to demonstrate that existing bioeconomy strategies reproduce the global social inequalities inherent in the production of knowledge. Proceeding from approaches inspired by world systems theory (WST) as well as post- and decolonial studies, I conceive of the global inequality in knowledge production not only as an expression of the unequal production of and access to technological research and development (R&D) on a global scale; rather, I seek to carve out the problematic notion of knowledge itself, as knowledge is inextricably linked to socio-ecological relations of power and inequalities. When I use the term "extractive knowledge" in this context, I am referring to R&D undertaken on behalf of agro-industrial resource extraction in (semi-)peripheral countries that is for the most part intended for export. As I explain below, this form of knowledge production has been part and parcel of socio-ecological inequality since colonial times.[1]

In the following, I begin by reconstructing the critical debate around the bioeconomy in the social sciences and demonstrate that the research standpoint proceeding from global inequality has so far been underrepresented in critiques of this new form of ecological modernisation. In Sect. 2.2, I outline my research perspective on global inequalities in knowledge production. Subsequently, I use the policy strategies put forward by the OECD, EU, Germany and Brazil to illustrate how the relations of inequality emanating from these papers are perpetuated. In Sect. 2.4, I draw on the case of Brazil to flesh out the socio-ecological implications that arise from the strengthening of extractive knowledge production and the simultaneous marginalisation of alternative knowledge and technologies. Although Brazil has yet to formulate

---

[1] By using the term 'socio-ecological', I emphasise the dialectical connection between nature and society. On this, see Görg (2004) and the Introduction to this volume.

a bioeconomy strategy, it is considered an important player in international bioeconomy forums. The country is one of the largest producers worldwide not only of biomass, but also of knowledge about genetically modified crops, the intensification of agro-industrial agriculture and the use of soybean and sugarcane for bioenergy. To conclude, I discuss the implications for the debate on a reorientation of the bioeconomy and the need for further research.

2.2 Bioeconomy and the Critique of This New Form of Ecological Modernisation

The bioeconomy is part of the green economy[2] and rests on the notion of ecological modernisation (Kleinschmit et al. 2014, p. 403). Proponents of ecological modernisation believe that economic growth can be decoupled from climate change or the overexploitation of natural resources. The preconditions for such a decoupling include, firstly, market-based ecological policies that set the right incentives for the private sector and consumers. A second central requirement is the funding of technological innovation aimed at facilitating a more efficient and environmentally friendly use of resources (Bemmann et al. 2014, p. 12; Mol et al. 2014). This notion of innovation is based on a specific understanding of knowledge and development. "Knowledge" does not refer to knowledge per se, but rather to that which is produced primarily by researchers in the natural sciences and eventually translated into innovations as part of a seemingly linear path of development. In the prevailing bioeconomic concept, innovation denotes the successful commercialisation of knowledge in the form of new products and services (Birch 2017, pp. 3–4). Given this technology-oriented development optimism, policymakers see no contradiction in the fact that current bioeconomy strategies also stipulate funding for R&D in conventional agriculture, even though it is

---

[2]The green economy was adopted as the guiding principle at the 2nd UN Summit on Sustainable Development Rio +20 in Rio de Janeiro in 2012. The fields of action go beyond the bioeconomy to include all technologies and political frameworks that contribute to the conservation of resources, sustainable consumption and mobility systems.

responsible for up to 30% of climate-damaging emissions (IPCC—Intergovernmental Panel on Climate Change 2019). Their aim is to render the agricultural regime more environmentally friendly and to make it climate neutral through the use of technologies aimed at increasing efficiency. At the same time, new sustainable fields of accumulation are to emerge as a result of the valorisation of technological innovation (particularly via patents) (Birch et al. 2010). Instead of recognising the "limits to growth", the aim is the "growth of the limits" (Escobar 1996, p. 330).

Countless articles criticise this orientation of the bioeconomy. At the heart of this critique is the technology optimism inherent to bioeconomy strategies, which result in government research funding benefiting primarily mainstream areas of (bio-)technology and agricultural research. One major problem is that the EU's policymakers reduce the ecological crisis to technological inefficiency (Birch et al. 2010). However, instead of eliminating socio-ecological problems, the numerous technological innovations that have been implemented over the past few decades with the aim of increasing the efficiency of agro-industrial and monocultural production in the agricultural sector have partly aggravated them (TNI and Hands on the Land 2015). Today, the globalised agro-industrial sector has negative impacts on working conditions as well as land access and land use for smallholders in production regions worldwide.[3] The growing demand for biomass in the context of bioeconomic policy (Bringezu et al. 2020; see also Lühmann in this volume) threatens to further entrench these social relations and exacerbate the ecological crisis, such as through the excessive consumption of freshwater and the emission of climate-damaging gases as a result of land use conversion or the growing use of fertilisers and agro-toxins (Fatheuer 2019; Moreno 2017). Adding to this is the increasing competition over the actual use of biomass either for food, energy or biochemical processing (ibid.).

Many critics, therefore, consider genuine socio-ecological and just solutions to be inconceivable within the dominant agricultural regime. They instead point to agroecology as an alternative approach to agricultural knowledge production (TNI and Hands on the Land 2015), which,

---

[3] See the articles by Lorenzen, Puder, Sinaga as well as Toledo-López in this volume.

if at all, has only ever been discussed on the margins or promoted in the context of bioeconomy strategies (Bugge et al. 2016; Diedrich et al. 2011; Levidow et al. 2012; Schmid et al. 2012). The public debates demanded by many critics about orienting research funding towards a socio-ecologically just bioeconomy, however, have yet to materialise (see Lehmann in this volume).

I broaden this critical perspective by including a global perspective on unequal knowledge production. After all, most studies on bioeconomic technology and knowledge production are regionally focused on Europe and North America.

2.3 Critical Perspectives on Unequal Global Knowledge Production

In order to develop a research perspective that proceeds from unequal global knowledge production, I propose a stronger focus on the notion of development that is inherent in the modernisation narrative, and which has been criticised by approaches rooted in WST and post- and decolonial theory. After all, ecological modernisation amounts to the continuation of the classical doctrine of modernisation (Bemmann et al. 2014), according to which, Western Europe and North America represent the role models of all societies throughout the world. From this perspective, the European notions of rationalism, science and technological progress are considered "the jewels in the crown of modernity" (Harding 2011, p. 2). Negative socio-ecological impacts of technological developments are said to have emerged as a result of the improper and inefficient application of technologies (ibid., p. 5). The Eurocentric understanding of science is based on the dichotomous dissociation from the "rest of the world" that is commonly denigrated as traditional (vs. modern), underdeveloped (vs. developed) and irrational (vs. rational) (Hall 1992). By divorcing science and technologies from their socio-historical embeddedness and universalising them, the modernisation narrative not only conceals the producedness of knowledge, but also renders the global relations of power and social inequalities, which permeate this knowledge, invisible and naturalises them (Harding 2011).

In the eyes of the proponents of ecological modernisation, it is therefore no contradiction that the old capitalist centres regard themselves as the trailblazers of environmental and climate protection.

Some important points of departure for my research are provided by studies from the WST research field. This analytical perspective (Wallerstein 2007) emphasises the significance of colonialism for the emergence of capitalism and focuses on the global inequalities that developed as a result. The unit of investigation encompasses the entire world system in which nation states are integrated according to a particular division of labour and global hierarchy as a consequence of colonialism. As centres, peripheries and semi-peripheries, they assume distinct positions within the relations of exchange in the world market. This historical perspective helps explain the relatively stable global hierarchies among nation states as well as the changes in their respective positions (e.g. China).

One important extension of WST is the research into unequal ecological exchange, which sheds light on how the global division of labour has been linked to the unequal extraction of resources and the distribution of environmental risks ever since colonial times (Bunker 1984; Gellert 2019).[4] Consequently, the major economic growth in the capitalist centres to this day would be inconceivable without the resource influx from the (semi-)periphery; at the same time, the (semi-)periphery acts as a sink for the outsourcing of environmentally harmful production (Lipke 2011, p. 351). These global socio-ecological inequalities are linked to asymmetrical political power: the proposed solutions for dealing with the ecological crisis continue to be dominated by the ideas of North American and Western European institutions (ibid.).

These power asymmetries are also interlinked with unequal scientific and technological knowledge production on a global scale. Another extension of WST are studies on the unequal global production of knowledge, science and technology as generated and dominated by the Western centres since colonial times (Demeter 2019). Although the main centres of scientific and technological knowledge production shifted from Western Europe to North America over the course of the twentieth

---

[4] On this, see Lühmann in this volume.

century, while East Asia became a more central region, the Latin American, Asian and African world regions have remained (semi-)peripheral (Schott 1998). This unequal global knowledge production is reproduced and, in part, even exacerbated, by the present-day political economy of science and academia (Demeter 2019). Empirical studies on global relations between centre, semi-periphery and periphery from the global natural sciences (Schott 1998), social sciences (Demeter 2019) and agricultural sciences (Delvenne and Kreimer 2017) consistently show the same tendency, albeit to varying degrees: despite the growing importance of China and other emerging countries in certain specialist fields (e.g. nanotechnology in China, biotechnologies in South Korea, or genetically modified crops in Brazil), the most influential journals as well as most publications and patents continue to come from the United States and Western Europe. Likewise, transnational research networks also reproduce these global inequalities in knowledge production, as a study of the EU research funding programme HORIZON suggests (Delvenne and Kreimer 2017, p. 394). According to the study's findings, these transnational research networks are for the most part headed by researchers and research institutions from the Western European and North American centres, whereas researchers from the (semi-)peripheries act as assistants and their share in theory formation only amounts to about 10% (ibid.). That said, global inequalities in knowledge production are not only perpetuated between countries, but also *within* countries. After all, whether they are in the centres or the peripheries, the only individuals who can successfully participate in "global" (or, more precisely, Anglo-American) science and academia are those who have gained the required professional experience abroad, the language skills they need, and whose class background provides them access to international networks (Demeter 2019).

The analytical perspectives on the unequal global production of knowledge as adopted by WST are conducive to comprehending the political economy of today's academic-scientific framework and the related global inequalities in knowledge production between and within countries. In the corresponding approaches, however, technological knowledge production *itself* does not take centre stage, as would be the case in research from a post- or decolonial perspective (Harding 2011;

Graddy-Lovelace 2016). In these perspectives, agro-industrial knowledge production, in the form that appeared at the latest with the emergence of the Green Revolution during the 1970s, is a modernisation project that started in the US and has come to dominate worldwide agriculture due to the work of international development organisations, and that is inextricably linked to continuing colonial power asymmetries and hierarchies (Harding 2011). Proceeding from approaches that focus on unequal ecological exchange, I thus propose referring to "extractive knowledge" in this context. I view extractive knowledge as agro-technological knowledge that allows for and reproduces unequal resource extractivism[5] as is manifest in globalised agriculture, in addition to all of the negative socio-ecological impacts and inequalities that are associated with it in production regions. This extractive knowledge, as a component of the agro-industrial sector, lies at the heart of the criticism of the orientation of research funding for bioeconomy strategies (see above). Despite the ascertained global social inequalities and power asymmetries, I conceptualise extractive knowledge not as monolithic, but as socially produced, historically situated and generally contested. After all, agro-industrial technologies are being challenged by agroecologists, social movements and NGOs worldwide.

Against this backdrop, I explore the following questions in a global context: Which and whose knowledge is to be funded in the context of the bioeconomy? Are global inequalities in extractive knowledge production challenged or reproduced? I answer these questions in two steps: first, I examine the bioeconomy strategy papers put forward by the EU, Germany, the OECD and Brazil. In a second step, I focus on Brazil's research funding in the agricultural sector, using the example of soybean.

---

[5]The term extractivism originally only referred to mining, but is now also used in the context of agriculture. For more details, see Tittor in this volume.

2.4 The Continued Global Division of Labour in Knowledge Production

An analysis of the bioeconomy strategy papers put forward by the OECD, the EU and Germany illustrate that they reproduce the inequalities in global knowledge production outlined above. This is evidenced by the fact that the new narrative of the bioeconomy as a green, market-based vision of the future was originally conceptualised by the OECD and is being substantively shaped and globally disseminated above all by the capitalist centres—especially the EU (Backhouse et al. 2017). As a result, the global policy for mitigating the ecological crisis—including with regard to the bioeconomy—is once again being defined by the capitalist centres.

In addition, these visions of a bioeconomy reproduce the global division of labour in technological knowledge production. The OECD and the EU locate the technology centres of this knowledge-based bioeconomy primarily in North America and Western Europe (OECD 2009; European Commission 2012, 2018). Similar to the EU as a whole (European Commission 2018, p. 4), Germany also emphasises its "global responsibility" (BMBF and BMEL[6] 2020, p. 4) as well as the need to initiate global research collaborations in order to generate synergies through an exchange of knowledge (ibid., pp. 33–34). This notion rests on the conviction that "each country and region can make an individual contribution to the global bioeconomy through its own mix of raw materials, technologies, knowledge and ideas" (ibid., p. 34). However, both papers are dominated by a competition-oriented perspective: it is all about maintaining and enhancing "its global leadership" in the development and marketisation of (bio-)technologies (European Commission 2018, p. 6; BMBF and BMEL 2020, p. 4). Correspondingly, the specified objectives do not foresee (semi-)peripheral countries rising above their role as raw material suppliers within the global division of labour. However, the OECD paper does consider changes in unequal globalised

---

[6] Federal Ministry of Education and Research (BMBF) and Federal Ministry of Food and Agriculture (BMEL).

knowledge production to be feasible and even describes them as a desirable outcome (OECD 2009, p. 198). Countries such as India, Brazil and especially China are viewed as having research capacities that are or could become relevant for the bioeconomy (ibid.). And yet, these papers do not address the unequal global distribution of the "high technological fields" (Delvenne and Kreimer 2017, p. 391) in the centres, and less complex technology domains in the (semi-)peripheries.

This modernisation narrative is being adopted by powerful players in industry, agriculture and politics in Brazil. According to the Brazilian Ministry of Science, Technology and Communications (MCTIC), the objective is to secure Brazil's leading role in the global competition for new technologies and markets by promoting and funding a higher degree of professional qualification and the expansion of its own biotechnological research in the fields of medicine, biotechnological industries and agriculture via government funding and public-private partnerships (MCTIC 2016). In this sense, a sub-chapter of the Brazilian research strategy demonstrates that Brazil's approach to the bioeconomy is similar to the approach put forward by the OECD in its 2009 strategy paper (see OECD 2009), namely in relation to the generation of biotechnological knowledge (MCTIC 2016, p. 96). In the global competition over pioneering biotechnology in the knowledge-based bioeconomy, Brazil believes its "comparative advantage" vis-à-vis other countries is the great wealth of biodiversity found in Brazil and the leading role the country plays in agribusiness[7] and biofuels (ibid., p. 96, own translation). A similar argument is put forward by the Brazilian National Confederation of Industry (CNI), which advocates a national bioeconomy strategy for Brazil and sets out the corresponding objectives and fields of action in its own papers (CNI 2014; Harvard Business Review 2013). In this case, sustainability and the reduction of CO_2 emissions in agriculture are to be achieved through biotechnological innovation (e.g. genetic modified organisms, synthetic biology) in the fields of bioenergy (sugarcane- and corn-based ethanol; soy-based biodiesel) and agriculture (e.g. new plant varieties such as eucalyptus, soy and corn) (Harvard Business Review 2013, pp. 19–20). The development goal—apart from exporting raw

---

[7] On the adoption of the US concept of agribusiness in Brazil, see Pompeia (2020).

materials—is the manufacture of high-quality and innovative products for the international market (ibid.).

The extent to which Brazil will actually manage to overcome global unequal knowledge production remains to be seen. Even if the research strategy of the former centre-left government of Dilma Rousseff is officially continued (MCTIC 2016; Mourão 2020), research policy has changed profoundly since radical right-wing President Jair Messias Bolsonaro took office in 2019. Universities and research centres in Brazil have suffered massive funding cuts.[8] As I demonstrate in the following section, the Brazilian adoption of the bioeconomy fosters a type of knowledge production that has been driving resource extraction since the Green Revolution of the 1970s—regardless of whether the respective (bio-)technologies have been imported or homegrown.

2.5 Extractive Knowledge Production in Brazil

As semi-peripheral country,[9] Brazil is by no means a mere recipient of agricultural technology from the North American and Western European knowledge centres. Brazil has ranked as the country with the largest agricultural research sector in Latin America since the year 2000 both in terms of research funding and the number of researchers with a PhD in this field (ASTI[10] 2016; IAASTD[11] 2009). According to a World Bank study from 2014, the Brazilian government invested about 1% of its GDP in agricultural research (by comparison: the US invested 1.4% of its GDP in the same year; see Correa and Schmidt 2014). A substantial share of these funds went to the Brazilian Agricultural Research Corporation (EMBRAPA) (ibid.), on which I focus in the following section.

---

[8] Public funding for science and technology was reduced from 6.37 billion reais in 2019 to 2.91 reais in 2020. See, http://www.portaltransparencia.gov.br/funcoes/19-ciencia-e-tecnologia?ano=2018. Accessed 20 Aug 2020.
[9] On the definition of semi-peripheral countries, see Delvenne and Kreimer (2017, p. 391).
[10] ASTI—Agricultural Science and Technology Indicators.
[11] IAASTD—International Assessment of Agricultural Knowledge, Science and Technology for Development.

Although EMBRAPA has recently suffered a loss to its erstwhile influence,[12] since its inception during the military dictatorship in the 1970s, the institute has significantly contributed to orientating the Brazilian agricultural sector towards the US model (Mengel 2015). As I would like to briefly outline using the example of soybean, this research orientation is an expression of extractive knowledge production that results in the use of new technological means to reproduce the unequal ecological exchange between the regions in Brazil that cultivate soybean and the recipient regions in Europe and China, and the Brazilian centres.

Alongside market liberalisation and expired patents, which reduced the price of agro-industrial inputs (herbicides, pesticides, fertilisers, machinery, etc.), as of the 1990s, it was the EMBRAPA's development departments that significantly contributed to the proliferation of soybean production in the savannah of Cerrado and the Amazon region, areas previously deemed inadequate for farming due to poor soil quality or climatic conditions. In order to do so, they promoted the use of genetically modified crops, improved farming methods such as no-till farming, and agrochemicals (Correa and Schmidt 2014).

Today, soy farming covers vast areas: the US Department of Agriculture (USDA) estimates that the soybean harvest in 2020/2021, which encompasses 38.5 million hectares (ha) of land (for comparison: the total area of Germany is 35.7 million ha), will yield 129 million metric tons (t) (USDA 2020). Brazil is thus not only the biggest producer of soybean in the world, but also the second-largest producer of "biotech crops", which includes genetically modified soybean, corn and cotton (ibid.). Some 96% of soybean produced in Brazil is genetically modified. In 2019 alone, applications for the commercial cultivation of 107 genetically modified crops were submitted in Brazil, 19 of which were varieties of soybean (ibid.). Exports of soybean are forecast at 79 million t for 2020/2021. China is the largest importer of Brazilian soybean (up to 75% of the Brazilian crop), followed by Europe (USDA 2019).[13]

---

[12] Under the Bolsonaro administration, funding for EMBRAPA was cut from 3.35 billion reais in 2019 to 1.75 billion reais in 2020. See, http://www.portaltransparencia.gov.br/orgaos/22202?ano=2020. Accessed 20 Aug 2020.

[13] On the restructuring of global inequalities and South-South relations, see Rodríguez in this volume.

The sector was additionally bolstered by the introduction of biodiesel in Brazil in 2004. Biodiesel consists of about 70% soybean oil, a waste product from the animal feed industry (USDA 2020). Since 2019, the blending quota has been at 11% and was raised to 12% in March 2020 (ibid.). However, corn-based[14] ethanol production (blending quota: 27%) is becoming increasingly appealing for the sector in soybean expansion regions, given that corn can be cultivated in crop rotation after harvesting soy (USDA 2019). Furthermore, the waste from corn ethanol production can be used as animal feed, which encourages the expansion of grazing pastures for cattle in the major growing regions located, in among other areas, in the Amazon region. Corn ethanol may thus "become a crucial connecting link in a new agro-industrial complex comprised of a combination of soybean, corn, ethanol and livestock farming" (Fatheuer 2019, p. 15; own translation), which could, in turn, reinforce and accelerate the dynamics of deforestation. Thus, technological innovations like crop rotation and cascade use of biomass alone are no guarantee of sustainable development.

The relationship between this form of knowledge production and socio-ecological inequalities in the Brazilian agricultural sector is evident when we consider the question of land access: the unprecedented expansion of soybean cultivation over the past 50 years has been made possible not only through technological innovation, but also because land access is controlled by a small elite in concert with major political actors. The massive concentration of land access has its origins in colonial times and represents one of the main historical reasons for Brazil's striking social inequalities. The expansion of soybean cultivation exacerbates these socio-ecological inequalities: the Federal State of Mato Grosso is one of the main growing regions, and around 80% of its farmland is owned by large landowners who own more than 1000 ha.[15] Soybean cultivation and the corresponding infrastructure (roads, ports) have become one of the main drivers of deforestation and the displacement of indigenous

---

[14]This is a new development. Until now, Brazilian ethanol has been based almost exclusively on sugarcane, see Backhouse (2020).
[15]Own calculation based on the 2017 agricultural census, see, https://sidra.ibge.gov.br. Accessed 20 Oct 2020.

people and (traditional) smallholders in the Amazon region today (Torres and Branford 2018).

Brazil's specialised knowledge production, therefore, not only limits agricultural research to the needs of agribusiness but, for decades, has also contributed to a fundamental restructuring of the Brazilian agricultural sector.

Furthermore, government programmes such as "Low-Carbon Emissions Agriculture" (*Agricultura de Baixa Emissão de Carbono*), which promotes technical efficiency increases via research funding for ecological modernisation, have also had no impact on the socio-ecological effects of soybean cultivation (Assad 2013). Technological innovations like precision agriculture, no-till farming, yield increases through new varieties, or rotating corn and soybean cultivation have failed to limit the socio-ecologically problematic expansion of cultivation areas. Instead, extractive knowledge production has been merely re-framed as part of ecological modernisation.

Alternative agroecological or traditional forms of agriculture, which are not incorporated into agribusiness (e.g. by contract farming), are not only displaced spatially, but also marginalised in terms of knowledge production. The major imbalances in the relations of power and social inequalities, as manifested spatially in the high degree of land ownership concentration in the hands of a small elite and politically in the influence of the agricultural lobby in parliament (*bancada ruralista*), also extend into the area of R&D. At the same time, this specialisation in knowledge production reinforces extraction relations due to the export orientation of the entire soybean sector (Backhouse and Lühmann 2020). Brazil's "individual contribution" to the global bioeconomy is thus associated with major socio-ecological problems at the local level with regard to both raw materials and knowledge.

Of course, the exclusion of alternative forms of knowledge in the agricultural sector does not mean that they do not exist or do not matter at all. The landless movement MST (*Movimento dos Trabalhadores Rurais sem Terra*) and other social movements have advocated agroecology and food sovereignty for many years. In the process, they have established their own knowledge centres such as the MST education centre *Escola Nacional Florestan Fernandes* and the agricultural engineering institute for land reform research Iterra (*Instituto Técnico de Pesquisa e Reforma Agrária*) in São Paulo. So far, these civil society actors have not participated in the (inter-)national bioeconomy forums and are, correspondingly, excluded from negotiations on the orientation of bioeconomy strategies.

2.6 Conclusion

As shown in this chapter, today's bioeconomy reproduces the existing global socio-ecological inequalities in the area of knowledge production. The technological leadership role of North America and Western Europe is already expressed in the corresponding policy papers. The (semi-)periphery participates in the bioeconomy through its respective fields of specialisation and expertise.

As previous critical analyses of these policy strategies have shown, and the case of Brazil underscores, the ecological modernisation narrative associated with the bioeconomy in agriculture is dominated primarily by the agro-industrial sector. There is a great danger that this research funding orientation will deepen the extractive relations between countries that export raw materials and those that consume them—at the same time, exacerbating the socio-ecological crisis. Research is still needed into the question of whether the current changes in research funding under the Bolsonaro administration are exacerbating these dynamics.

The case of Brazilian soybean underlines the point that improved technologies may provide yield increases, but this will not necessarily stop the expansion dynamics associated with a particular crop. A decrease in biodiversity and unequal relations of access to and use of land can even be compounded as a result. Any meaningful socio-ecological solution within this agro-industrial agricultural regime can thus be ruled out. This makes it all the more important to develop alternative visions to this extractive knowledge. The great challenge that remains for critical actors in academia and science as well as in civil society, therefore, is to re-politicise and democratise not only the field of debate surrounding the bioeconomy, but technological knowledge production and government research funding as a whole. This is the only way to help initiate social negotiation processes that would further the development of alternative bioeconomies in highly distinct contexts and countries. Starting points for democratic fields of experimentation of alternative knowledge production that merit public funding, including social movements for agroecology and food sovereignty, can be found worldwide.

References

Assad, E.D. (2013). *Agricultura de Baixa Emissão de Carbono: A evolução de um novo paradigma*. São Paulo: Fundação Getulio Vargas; Centro de Agronegócio, Escola de Economia de São Paulo.

ASTI (2016). Agricultural R&D Indicators Factsheet: Brazil. Key indicators. https://www.asti.cgiar.org/pdf/factsheets/Brazil-Factsheet.pdf. Accessed 20 Aug 2020.

Backhouse, M. (2020). The Knowledge-Based Bioeconomy in the Semi-Periphery: A Case Study on Second-Generation Ethanol in Brazil. *Working Paper 13, Bioeconomy & Inequalities*, Jena. https://www.bioinequalities.uni-jena.de/sozbemedia/wp/workingpaper13.pdf. Accessed 20 Aug 2020.

Backhouse, M., Lorenzen, K., Lühmann, M., Puder, J., Rodríguez, F., & Tittor, A. (2017). Bioökonomie-Strategien im Vergleich: Gemeinsamkeiten, Widersprüche und Leerstellen. *Working Paper 1, Bioeconomy & Inequalities*, Jena. https://www.bioinequalities.uni-jena.de/sozbemedia/neu/2017-09-28+workingpaper+1.pdf. Accessed 17 Dec 2018.

Backhouse, M., & Lühmann, M. (2020). Stoffströme und Wissensproduktion in der globalen Bioökonomie: Die Fortsetzung globaler Ungleichheiten. *Peripherie 159/160*, 235–257.

Bemmann, M., Metzger, B., & von Detten, R. (2014). Einleitung. In B. Metzger, M. Bemmann & R. von Detten (Eds.), *Ökologische Modernisierung: Zur Geschichte und Gegenwart eines Konzepts in Umweltpolitik und Sozialwissenschaften* (pp. 7–32). Frankfurt: Campus Verlag.

Birch, K. (2017). *Innovation, Regional Development and the Life Science: Beyond Clusters* (Regions and cities, 105). London, New York: Routledge Taylor & Francis Group.

Birch, K., Levidow, L., & Papaioannou, T. (2010). Sustainable Capital? The Neoliberalization of Nature and Knowledge in the European "Knowledge-based Bio-economy". *Sustainability, 2*, 2898–2918.

BMBF & BMEL (2020). *National Bioeconomy Strategy*. Berlin. https://biooekonomie.de/sites/default/files/bmbf_national-bioeconomy-strategy_en_0.pdf. Accessed 19 Oct 2020.

Bringezu, S., Banse, M., Ahmann, L., Bezama, A., Billig, E., Bischof, R., et al. (2020). Pilotbericht zum Monitoring der deutschen Bioökonomie. Kassel University, Center for Environmental Systems Research (CESR). https://kobra.uni-kassel.de/handle/123456789/11591. Accessed 28 Sep 2020.

Bugge, M., Hansen, T., & Klitkou, A. (2016). What Is the Bioeconomy? A Review of the Literature. *Sustainability, 8*, 691.

Bunker, S.G. (1984). Modes of Extraction, Unequal Exchange, and the Progressive Underdevelopment of an Extreme Periphery: The Brazilian Amazon, 1600-1980. *American Journal of Sociology, 89*(5), 1017–1064.

CNI (2014). Bioeconomia: oportunidades, obstáculos e agenda: Mapa Estratégico da Indústria 2013-2022. Uma agenda para a competitividade. Brasília. http://arquivos.portaldaindustria.com.br/app/conteudo_24/2014/07/22/479/V35_Bioeconomiaoportuidadesobstaculoseagenda_web.pdf. Accessed 27 Aug 2020.

Correa, P., & Schmidt, C. (2014). Public Research Organizations and Agricultural Development in Brazil: How Did Embrapa Get It Right? Economic Premise, 145. The World Bank, 1–10. http://documents1.worldbank.org/curated/en/156191468236982040/pdf/884900BRI0EP1450Box385225B000PUBLIC0.pdf. Accessed 9 March 2020.

Delvenne, P., & Kreimer, P. (2017). World-System Analysis 2.0: Globalized Science in Centers and Peripheries. In D. Tyfield, C. Thorpe, R. Lave & S. Randalls (Eds.), *The Routledge Handbook of the Political Economy of Science* (pp. 390–404). London: Taylor & Francis.

Demeter, M. (2019). The World-Systemic Dynamics of Knowledge Production: The Distribution of Transnational Academic Capital in the Social Sciences. *Journal of World-Systems Research, 25*, 111–144.

Diedrich, A., Upham, P., Levidow, L., & van den Hove, S. (2011). Framing Environmental Sustainability Challenges for Research and Innovation in European Policy Agendas. *Environmental Science & Policy, 14*, 935–939.

Escobar, A. (1996). ConstructionNature: Elements for a Post-Structuralist Political Ecology. *Futures, 28*(4), 325–343.

European Commission (2012). *Innovation for Sustainable Growth: A Bioeconomy for Europe: Communication from the Commission to the European Parliament, the Council, The European Economic and Social Committee and the Committee of the Regions*. Brussels. http://ec.europa.eu/research/bioeconomy/pdf/official-strategy_en.pdf. Accessed 5 June 2015.

European Commission (2018). *A Sustainable Bioeconomy for Europe: Strengthening the Connection Between Economy, Society and the Environment*. Brussels. https://ec.europa.eu/research/bioeconomy/pdf/ec_bioeconomy_strategy_2018.pdf. Accessed 17 May 2020.

Fatheuer, T. (2019). Zuckerträume: Ethanol aus Brasilien in der globalen Klimapolitik. FDCL. Berlin. https://www.fdcl.org/wp-content/uploads/2020/03/FDCL_Zuckertra%CC%88ume_web.pdf. Accessed 23 June 2020.

Gellert, P.K. (2019). Bunker's Ecologically Unequal Exchange, Foster's Metabolic Rift, and Moore's World-Ecology: Distinctions With or Without a Difference? In R.S. Frey, P.K. Gellert & H.F. Dahms (Eds.), *Ecologically Unequal Exchange* (pp. 107–140). Cham: Palgrave Macmillan.

Görg, C. (2004). The Construction of Societal Relationships with Nature. *Poiesis & Praxis*, 3(1), 22–36.

Graddy-Lovelace, G. (2016). The Coloniality of US Agricultural Policy: Articulating Agrarian (In)Justice. *Journal of Peasant Studies*, 44(1), 78–99.

Hall, S. (1992). The West and the Rest: Discourse and Power. In S. Hall & B. Gieben (Eds.), *Formations of Modernity* (pp. 185–227). Cambridge: Polity Press in association with Blackwell and the Open University.

Haraway, D. (1988). Situated Knowledges: The Science Question in Feminism and the Privilege of Partial Perspective. *Feminist Studies*, 14(3), 575–599.

Harding, S.G. (2011). Introduction: Beyond Postcolonial Theory: Two Undertheorized Perspectives on Science and Technology. In S.G. Harding (Ed.), *The Postcolonial Science and Technology Studies Reader* (pp. 1–31). Durham: Duke University Press.

Harvard Business Review (2013). Bioeconomy: An Agenda for Brazil. http://arquivos.portaldaindustria.com.br/app/conteudo_24/2013/10/18/411/20131018135824537392u.pdf. Accessed 19 Sep 2017.

IAASTD (2009). Agriculture at a Crossroad: International Assessment of Agricultural Knowledge, Science and Technology for Development. Synthesis Report. Washington, D.C. https://www.weltagrarbericht.de/fileadmin/files/weltagrarbericht/IAASTDBerichte/SynthesisReport.pdf. Accessed 19 June 2020.

IPCC (2019). IPCC Special Report on Climate Change, Desertification, Land Degradation, Sustainable Land Management, Food Security, and Greenhouse Gas Fluxes in Terrestrial Ecosystems: Summary for Policymakers Approved Draft. https://www.ipcc.ch/site/assets/uploads/2019/08/Edited-SPM_Approved_Microsite_FINAL.pdf. Accessed 7 April 2020.

Kleinschmit, D., Lindstad, B.H., Thorsen, B.J., Toppinen, A., Roos, A., & Baardsen, S. (2014). Shades of Green: A Social Scientific View on Bioeconomy in the Forest Sector. *Scandinavian Journal of Forest Research, 29*, 402–410.

Levidow, L., Birch, K., & Papaioannou, T. (2012). Divergent Paradigms of European Agro-Food Innovation: The Knowledge-Based Bio-Economy (KBBE) as an R&D Agenda. *Science, Technology, & Human Values, 38*(1), 94–125.

Lipke, J. (2011). Globale Herrschaftsverhältnisse und Naturaneignung: Eine weltsystemische und sozial-ökologische Betrachtung der globalen Umweltkrise. In T. Mayer, R. Meyer, L. Miliopoulos, P.H. Ohly & E. Weede (Eds.), *Globalisierung im Fokus von Politik, Wirtschaft, Gesellschaft: Eine Bestandsaufnahme* (pp. 351–371). Wiesbaden: VS Verlag für Sozialwissenschaften.

MCTIC (2016). Estratégia Nacional de Ciência, Tecnologia e Inovação 2016-2022. Brasília. http://www.finep.gov.br/images/a-finep/Politica/16_03_2018_Estrategia_Nacional_de_Ciencia_Tecnologia_e_Inovacao_2016_2022.pdf. Accessed 23 Aug 2020.

Mengel, A.A. (2015). *Modernização da agricultura e pesquisa no Brasil: a Empresa Brasileira de Pesquisa Agropecuária – EMBRAPA*. PhD. Universidade Federal Rural do Rio de Janeiro, Rio de Janeiro.

Mol, A.P.J., Spaargaren, G., & Sonnenfeld, D.A. (2014). Ecological Modernisation Theory: Where Do We Stand? In B. Metzger, M. Bemmann & R. von Detten (Eds.), *Ökologische Modernisierung: Zur Geschichte und Gegenwart eines Konzepts in Umweltpolitik und Sozialwissenschaften* (pp. 35–66). Frankfurt: Campus Verlag.

Mourão, A.H.M. (2020): Amazônia: a Nova Fronteira da Bioeconomia. *Interesse Nacional, 1*, 9–13.

Moreno, C. (2017): Landscaping a Biofuture in Latin America. FDCL. Berlin. https://www.fdcl.org/wp-content/uploads/2017/07/FDCL_BIOEC_EN18072017-2.pdf. Accessed 20 Aug 2020.

OECD (2009). *The Bioeconomy to 2030: Designing a Policy Agenda*. Paris. http://biotech2030.ru/wp-content/uploads/docs/int/The%20Bioeconomy%20to%202030_OECD.pdf. Accessed 5 June 2020.

Pompeia, C. (2020). "Agro é tudo": simulações no aparato de legitimação do agronegócio *Horizontes Antropológicos, 26*(56), 195–224.

Schmid, O., Padel, S., & Levidow, L. (2012). The Bio-Economy Concept and Knowledge Base in a Public Goods and Farmer Perspective. *Bio-based and Applied Economics, 1*(1), 47–63.

Schott, T. (1998). Ties between Center and Periphery in the Scientific World-System: Accumulation of Rewards, Dominance and Self-Reliance in the Center. *Journal of World-Systems Research, 4*(2), 112–144.

TNI & Hands on the Land (2015). The Bioeconomy. A Primer. https://www.tni.org/files/publication-downloads/tni_primer_the_bioeconomy.pdf. Accessed 1 Sep 2018.

Torres, M., & Branford, S. (2018). *Amazon Besieged: By Dams, Soya, Agribusiness and Land-Grabbing*. Warwickshire, UK: Practical Action Publishing.

USDA (2019). Brazil: Oilseeds and Products Annual. GAIN Report (BR1906). https://apps.fas.usda.gov/newgainapi/api/report/downloadreportbyfilename?filename=Oilseeds%20and%20Products%20Annual_Brasilia_Brazil_4-2-2019.pdf. Accessed 19 June 2020.

USDA (2020). Oilseeds and Products Annual: Brazil (BR2020-0011). https://apps.fas.usda.gov/newgainapi/api/Report/DownloadReportByFileName?fileName=Oilseeds%20and%20Products%20Annual_Brasilia_Brazil_04-01-2020. Accessed 19 June 2020.

Wallerstein, I. (2007). *World-Systems Analysis: An Introduction* (5th ed., A John Hope Franklin Center book). Durham: Duke University Press.

Open Access This chapter is licensed under the terms of the Creative Commons Attribution 4.0 International License (http://creativecommons.org/licenses/by/4.0/), which permits use, sharing, adaptation, distribution and reproduction in any medium or format, as long as you give appropriate credit to the original author(s) and the source, provide a link to the Creative Commons license and indicate if changes were made.

The images or other third party material in this chapter are included in the chapter's Creative Commons license, unless indicated otherwise in a credit line to the material. If material is not included in the chapter's Creative Commons license and your intended use is not permitted by statutory regulation or exceeds the permitted use, you will need to obtain permission directly from the copyright holder.

3

Neoliberal Bioeconomies? Co-constructing Markets and Natures

Kean Birch

3.1 Introduction

If we are to meet the targets of the 2016 Paris Agreement—especially, as stated in the document, 'to pursue efforts to limit the [global] temperature increase even further to 1.5 degrees Celsius' by 2030—then we need to do more, much more, and do so now. Primarily, we need to find ways to transition our carbon economies and societies to a low-carbon future, and do so with some urgency. How we go about this transition is the real issue we face now. At points like this, I am always reminded of Bill McKibben's 2012 *Rolling Stone* article—'Global warming's terrifying math'—when it comes to the urgency of climate change: simply put, to keep to 2 degrees Celsius, he argued that humans can only release another 565 gigatons (Gt) of carbon dioxide. And that was back in 2012, we are now at somewhere around another 350 Gt.

K. Birch (✉)
Faculty of Environmental and Urban Change, York University, Toronto, ON, Canada
e-mail: kean@yorku.ca

Several low-carbon transition pathways have been suggested in response to this very urgent imperative. One such pathway is called the 'bioeconomy'. The bioeconomy—or, 'bioeconomies'—is premised on replacing fossil fuels with renewable biological materials (e.g. plants, algae etc.) as the key underpinning resource in our economies (European Commission 2012; The White House 2012; German Bioeconomy Council 2015a, b). It is usually presented as a market-based transition pathway, rather than a wholesale transformation of our societies or economies—although it is also portrayed as the latter by some (e.g. Schmid et al. 2012). The bioeconomy is supposed to be a more sustainable (capitalist) economy because it is based on renewable resources that produce fewer greenhouse gas (GHG) emissions over their industrial life-cycle (Birner 2018). As a transition pathway, then, the bioeconomy entails a specific material political economy in which markets and natures are co-constructed. While this might seem like a classic case of the *neoliberalization of nature* (Castree 2008a, b; Bakker 2009; Bigger and Dempsey 2018), a more complicated process is at play, as I have discussed elsewhere (Birch 2019).

The bioeconomy is often presented as a 'business-as-usual' approach to resolving the problems of climate change, which tend then not to provide an actual solution to these problems (see Tyfield 2017). It has, as a result, been criticized for being too market-centric—or 'neoliberal'—by a number of people, including myself. Despite the value in this 'neoliberal natures' approach (e.g. Kenney-Lazar and Kay 2017), I have found that this neoliberal natures framing frequently closes down debate about the bioeconomy, leaving little room to develop alternative bioeconomy approaches (e.g. agroecology). In particular, the neoliberal natures literature tends to reduce the relationship between markets and natures to a problematic imposition of markets as a social aberration on romanticized natures, even presenting nature as contesting or fighting back against neoliberalism. In this chapter, my aim is to problematize this neoliberal framing of the bioeconomy by exploring the co-construction of markets and natures, rather than the imposition of one on the other. I start by outlining what I mean by neoliberalism and neoliberal natures; I do so in order to emphasize the particularities and limitations of this approach. I then discuss the co-construction of markets and natures in the bioeconomy as a way to try and understand 'neoliberal bioeconomies'.

3.2 Neoliberalism

3.2.1 What Is Neoliberalism?

Neoliberalism is a term usually used to critique the prevailing *market-*based logics and responses to climate change, acknowledging that the term is increasingly contested (Springer et al. 2016; Birch 2017). It has been used in various ways over the last few decades, which means it can be difficult to parse what is meant when scholars—or others—use the concept. However, it is commonly used—across different critical traditions—as a way to characterize the expansion and extension of markets as the main way to organize society—across several different neoliberal schools of thought. In this sense, it is very much a political and analytical term, since it is frequently used to refer to a particular 'market ethic' (Harvey 2005) in which liberty and freedom are assumed to arise from private property rights and market contracts—see, for example, the arguments of people like Hayek (2001, 2011) or Friedman (1962). As mentioned, there are a number of schools of neoliberalism and numerous analytical traditions that are critical of these neoliberal schools and their ideas. I can only briefly discuss some of the differences here, before outlining how the extension of markets to environmental issues has been criticized.

It is possible to identify different schools of neoliberal thinking. The most well-known include the Austrian, Freiburg/Ordoliberal, Chicago, and Virginia schools (Birch 2015). However, when most people write about neoliberalism nowadays, they generally erase the nuances between differing schools by associating neoliberalism with the (later) Chicago and/or Virginia Schools. These two schools assume that everything can be treated as a market because they conceptualize everything as already a market (see Amadae 2016). The effects of this are to naturalize markets, thereby legitimating the installation of markets everywhere and the removal of state intervention in a naturalized 'free' market (Birch 2017).

The critical analytical traditions that have arisen in response to these schools of neoliberal thought are also pretty varied. Each critical tradition is different, but they share one commonality: the idea that neoliberalism entails the spread and entrenchment of markets (or market proxies) across society.

First, one of the earliest analysts of neoliberalism was Michel Foucault (2008), whose lectures on *The Birth of Biopolitics*, held between 1978 and 1979, provide the groundwork for a lot of later scholarship. In particular, Foucault outlined two modern variants of liberalism—Chicago neoliberalism and Ordoliberalism—that share similar political rationalities while differing in terms of the technologies they deploy in the governing of national populations. Later work by Dardot and Laval (2014), amongst others, draws on these insights to update Foucault for the twenty-first century. Generally, they are concerned with how neoliberalism produces specific subjectivities, identities, social relations, and so on; these are largely configured by the 'economization' of social life through the construction of individuals as what I call *market monsters* (Birch 2017). Here, these modern Foucauldians stress the individual transformation into an 'entrepreneur of the self'—that is, the reconstruction of our selves through our acquiescence to a market (or business) logic in our ways of engaging with the world. We come, in this Foucauldian sense, to think *always* like a market.

Second, a similar tendency to subsume individual reflexivity under all-consuming market logics is also evident in the various Marxist— or Marxist-inflected—takes on neoliberalism. One set of perspectives frames neoliberalism as an elite class project, entailing the dispossession of our commons (e.g. nature) with an ideological worship of markets (e.g. Harvey 2005). Critically, this perspective acknowledges that elite interests often end up side-lining market-based rationales and legitimation where they come against the restoration of elite class power (ibid.). Class also figures in other Marxist perspectives, such as the state-theoretic approach of regulation thinkers, which has influenced much of the geographical and sociological literature on neoliberalism (Birch 2017). In particular, the geographical analysis of neoliberalism has tended to frame it as a 'process' of uneven political-economic restructuring—which generates messy and uncertain outcomes (e.g. Peck and

Tickell 2002; Larner 2003; Castree 2008a). One of the issues with this critical geographical take is that it rolls everything into the 'neoliberalization' process such that state, market, and non-governmental actors all end up implicated in the roll-out of markets—especially when it comes to resolving environmental problems, which I discuss below.

Finally, more recent critical literature on neoliberalism has tended towards ideational analyses of neoliberal concepts and their influence. Much of this tradition is based on philosophy, history, and political science. Key exponents of this view include Mirowski (2013) and his collaborators. They place greater emphasis on the power of ideas to shape material interests and political decision-making, especially through the creation of 'thought collectives' (ibid.). The epistemic tradition equates the spread of neoliberalism with—usually right-wing—political movements, meaning that they are concerned with how market-based logics are taken up.

3.2.2 Neoliberalizing Nature

While the above can only provide a brief introduction to the theoretical complexities of neoliberalism, it is helpful for introducing the key conceptual approach used to understand and critique the deployment of market-based instruments to solve environmental problems. Emerging over the last decade or so, this approach is generally defined by its focus on 'neoliberal natures' and has gradually built up a significant scholarship on a range of topics (see Bigger and Dempsey 2018). These topics, listed alphabetically, include agriculture (e.g. Essex 2016), biofuels (e.g. Birch et al. 2010; Levidow et al. 2012), climate change (Lohmann 2016), ecosystem services (e.g. Dempsey and Robertson 2012), forestry (e.g. Prudham 2005), genetics and genomics (e.g. McAfee 2003), and water (e.g. Loftus and Budds 2016). Several thorough reviews of this literature have also been produced over the last decade or so, including those by Castree (2008a, b, 2010a, b), Bakker (2009, 2010), and Collard et al. (2016).

Across this *neoliberal natures* literature, the proponents of market-based instruments and mechanisms are framed as advocates of certain political-economic processes as well as certain policies (e.g. carbon pricing and trading) and technological solutions (e.g. biotechnology) as ways to resolve environmental problems. Markets are meant to solve a range of environmental problems, including the over-use of commonly-held resources (e.g. wetlands, oceans) through the extension of private property rights; the externalities generated form industrial activity (e.g. pollution, fertilizer runoff) through the creation of new quasi-commodities like emissions credits; and declining ecological and agricultural productivity (e.g. crop growth, bee loss) through new technologies like genetically modified seeds. Here, the role of the state is framed as an advocate and supporter of market-based solutions, facilitating their roll-out, rather than as a political means for collective action. Much of the critique of these market-based instruments and mechanisms centres on an understanding of them as a process—that is, on the neoliberalization of nature. Scholars working in this critical field are concerned with the specificities of this neoliberalization process, especially with the changes caused by the privatization, commodification, and marketization of nature. For example, *privatization* represents a sale of public assets (e.g. forest) to private sector actors; it is similar to *dispossession* although the latter entails the wholesale transfer (cf. sale) of public assets to the private sector without monetary returns.

There are at least two aspects of these debates worth considering more critically when it comes to understanding the bioeconomy, to which I turn in the next section. Both relate to the analysis of the biophysical materialities of markets and nature—and they both problematize the critique underlying the idea of the neoliberalization of nature (Birch 2019).

First, part of the analytical value of the neoliberal natures literature is the promise of theoretical consistency across various strands of research and substantive topics, outlining precisely what 'neoliberalization' process applies to what 'nature' (Bakker 2009). Something like marketization, for example, should share analytical similarities in its use throughout this literature for it to make sense to use the term 'neoliberal' as a way to define various happenings. Castree (2008a, p. 142) defines marketization as 'the assignment of process to phenomena' where, it is important to stress, something was 'previously shielded from market exchange'. Later, Castree (2010b, p. 1728) defines marketization as 'rendering alienable and exchangeable things that might not previously have been subject to a market', which is similar but also slightly different from the earlier definition. Both, though, treat the 'market' as a given (i.e. already existing), rather than social construct or instituted process (Polanyi 2001 [1944]), a point I will come back to. More broadly, Bakker (2010, p. 723) defines marketization as when 'markets determine resource allocation and pricing'—again, markets are treated as given, although she pluralizes them. Both scholars treat marketization as a 'political-economic' transformation, as something that happens to or is imposed on environmental phenomena (which was not subject to pricing beforehand). Markets are treated as an alien imposition on a natural phenomenon, as outside a set of natural/naturalized processes (e.g. rivers, forests, etc.).

Second, the neoliberal natures literature tends to valorize nature/s and represent nature/s as 'resisting' or 'contesting' market mechanisms, instruments and logics—this includes a range of environmental processes or systems (e.g. trees that do not grow straight, making it more difficult to harvest them). Nature resists neoliberalism to many of these thinkers; it has an agential materiality. An example is Castree's (2010a, p. 1752) comments that neoliberalism is 'defined by its engagement with the non-human world' and the 'challenge' nature represents to 'neoliberal policies over time'. Other examples include McCarthy and Prudham (2004), who argue that nature represents a 'check' on neoliberalism; Fletcher (2014), who argues that natural 'recalcitrance' limits neoliberalism; and Roff (2008), who argues that nature represents a fundamental challenge to neoliberalism. Such contestation is framed as reconfiguring neoliberalism; for example, privatization of water is disrupted by its biophysical materialities (Bakker 2010). Across this neoliberal natures literature, then, markets are characterized as an aberration of nature— its antithesis. As such, it actually repeats and reinforces the notion that political-economic and natural processes are distinct from one another, whether that is the intent or not. In a way, it naturalizes the idea that our biophysical world is the starting condition *on which* we end up acting.

Elsewhere, I have sought to push against these analytical assumptions, or starting points (Birch 2019). I think it is important to problematize the idea that material nature is transformed by social-economic processes, on the one hand, and that markets are aberrations of a pristine nature, on the other hand. In contrast, I have sought to analyse the co-construction of markets and natures in order to understand the entanglement of our political-economic artefacts (e.g. markets) and biophysical materialities (e.g. nature). My point here is that nature and political economy are not distinct from one another. As Jason Moore (2015) notes, capitalism has an ecology to it. Markets and natures are co-constructed, meaning that specific markets emerge in conjuncture with specific natures. The question we need to ask then is what type of market-natures are we dealing with and how are they co-constructed.

3.3 Neoliberal Bioeconomy? Co-constructing Markets and Natures

As noted, a simple definition of the bioeconomy is the use of biomass (e.g. plants) as the main resource in the production of energy, goods and services, although this definition obscures the different emphases that different people place on it (see Birch 2019). The bioeconomy first emerges as a key policy strategy in the mid-2000s when both the OECD and European Commission (EC) produce policy visions and frameworks for its development. More recent policy strategies include those by the EU and the White House (e.g. European Commission 2012; The White House 2012). Although a rather esoteric concept—in that it was and still is rarely discussed outside of policy circles—the bioeconomy has become a major strategy in a growing number of countries (German Bioeconomy Council 2015a, b). There have been several reviews of the bioeconomy as a policy concept, strategy and framework, including work by myself (e.g. Birch et al. 2010; Birch 2016a, b, 2019). Others have stressed a range of dimensions to the bioeconomy, including its national and subnational characteristics (e.g. McCormick and Kautto 2013; Staffas et al. 2013), its relationship to sustainability (e.g. Pfau et al. 2014; El-Chickakli et al. 2016), its diverse manifestations and geographies (e.g. Bugge et al. 2016; Calvert et al. 2017b; Hausknost et al.

2017), and its political implications (e.g. Frow et al. 2009; Richardson 2012; Mukhtarov et al. 2017).

Rather than dwell on these aspects of the bioeconomy, though, my focus here is on how it has been implemented and how this entails the co-construction of specific markets and natures. As a potential transition pathway, then, the bioeconomy cannot be imposed top-down on an economy or natural environment as a simple policy proposal and policy framework. Its success necessarily depends on the configuration of a new political economy and a new natural environment, which can happen in different ways and involve different bioeconomies, some of which receive more policy support than others. As many authors note (e.g. Levidow et al. 2012; Schmid et al. 2012), bioeconomies can be very different from one another, and this impacts how we understand the bioeconomy and its potential. For example, a bioeconomy based on agroecology will involve a very different configuration of political economy and natural environment than one based on hi-tech biological technologies. The former has the potential to be more distributed, localized and democratic compared with the latter, which is determined more by centralization tendencies and capitalist imperatives. Evidently, these differences are important to study and analyse because they frame how we might want to roll-out the bioeconomy as a policy strategy and low-carbon transition pathway.

3.3.1 Market Development Policies for the Bioeconomy

To date, the bioeconomy has mostly been implemented through the roll-out of 'market development policies' (MDPs), driven by prevailing capitalist logics rather than challenging them. These MDPs are especially evident when it comes to the development of biofuels markets, which is the focus of the rest of this chapter. All such MDPs for biofuels are good examples of the way that markets are socially *instituted* and *organized* (à la Polanyi 2001 [1944]), rather than being some sort of naturalistic mechanism or set of economic laws. In examining these MDPs, it also becomes possible to see how their implementation is co-constructed with specific biophysical materialities. At their base, these MDPs include a range of policy actions, including subsidies to support research and development as well as pilot or demonstration projects; mandates to regulate supply and demand; standards to integrate sustainability criteria and measurement; and physical infrastructure to embed supply chains (see Daemmrich 2015 on bioplastics and Birch 2019 on advanced biofuels for examples). A range of MDPs have been implemented around the world (see Table 3.1).

As this chapter draws on empirical material from Canada on the development of markets for conventional and advanced biofuels, I am going to outline briefly some of the relevant MDPs implemented in the Canadian context. I draw on Birch (2016a, 2019), Birch and Calvert (2015), and Calvert et al. (2017a) as my main sources for the rest of this section. These MDPs cut across federal and provincial scales and include those discussed in Table 3.1 (e.g. biofuels mandates, subsidies, standards), as well as others not included (e.g. feedstock supply chains).

First, the Canadian federal government's *Renewable Fuel Regulations* (RFR)—enacted in 2006 and implemented in 2010/2011—is the main biofuels mandate, stipulating 5% renewable content by 2010 for petroleum and 2% for diesel by 2011. The RFR does not mandate advanced biofuels, unlike similar biofuels mandates in countries like the US. Second, Canada has and has had a range of subsidies for the development of conventional and advanced biofuels: it started with a tax exemption scheme that was phased out in 2008 and replaced with a production credit, largely as a way to support domestic producers since anyone could claim the earlier tax exemption. Other initiatives included support for building new refineries. Third, Canada put in place feedstock supply chains as a way to ensure continuous supply through long-term contractual arrangements, like long-term timber cutting leases. Finally, Canada has participated in the development of international biofuels standards (e.g. ISO/TC 28/SC 7 Liquid Biofuels), although there have been significant limits on whether these standards can incorporate non-technical elements (e.g. sustainability, environmental and social goals).

3 Neoliberal Bioeconomies? Co-constructing Markets and Natures

Table 3.1 Market development policies around the world

Policies	Details	Examples
Mandates	Covers biofuel blending mandates and renewable fuel standards (RFS) that require a particular percentage of biofuels in retail petroleum or specific volume of biofuel production	The US has a RFS stipulating the production of 136 billion litres of biofuels by 2022; the EU's 2009 Renewable Energy Directive (RED) set a 10% target for biofuels in transport fuels by 2020
Subsidies	Covers range of subsidies for bio-based products and energy, including biofuels. These subsidies range from incentives for energy production through funding for demonstration plants to loans and grant support for facility construction	Germany and the UK provide financial support for the development of demonstration and pilot plants
Research funding	Covers basic and applied research funding	Most countries have research support specifically directed at areas like biotechnology, biofuels, renewable energy, bio-based products, etc.; the EU, for example, has focused a significant proportion of Framework Programme 7 funding on the cross-cutting theme of the 'knowledge-based bio-economy'
Standards & certification	Covers the establishment of standards and the certification of new products and services, especially where this might involve the incorporation of sustainability criteria	The EU has established standards for bio-based products (e.g. CEN/TC 411)

(continued)

Table 3.1 (continued)

Policies	Details	Examples
Labelling	Covers the creation of labels designed to create greater consumer awareness of new products and their sustainability characteristics	France has established the label *batiment biosourcé* for bio-based buildings

Source Adapted from Birch (2019)

3.3.2 Co-construction of Markets and Natures in the Bioeconomy

While the MDPs outlined above provide some insight into the policy development of the bioeconomy, at least in relation to biofuels, they also only provide a one-sided take on the instituting of markets as a solution to environmental problems. In particular, focusing on MDPs in this way obscures the materialities of markets, by which I mean the ways that markets and the natures are co-constructed (Birch and Calvert 2015; Becker et al. 2016; Birch 2019).

Starting with *biomass availability* for the bioeconomy, it is evident that the bioeconomy is premised on more than the total amount of biomass available; the biophysical materialities of the biomass itself configure the bioeconomy. A considerable proportion of Canada's land, for example, is Crown Land (i.e. it is owned by the state), including land harvested for biomass (e.g. forests). Rights to harvest on Crown Land are leased on a long-term basis and the harvest covers a variety of tree species; access to those trees depends on the materialities of access to the biomass. In Ontario, for example, the development of advanced biofuels from forest biomass is only viable economically if biofuels developers do not have to build the physical infrastructure to access the biomass (e.g. forest roads); the Provincial Government, instead, builds and maintains forest access roads. This enables timber harvesting by holders of long-term forest licences—who can sell their licence to others—but it does nothing for private woodlot owners who lack the public support and funding to access their forest assets. As such, the political-economic materialities here actually limit market competition—contrasting with the arguments made in the neoliberal natures literature—and mean that the bioeconomy does not have to be subject to the same market pressures as emphasized in current scholarly debates on neoliberalism (e.g. Castree 2008a; Bigger and Dempsey 2018). Rather, biomass availability is constituted by an interplay between the biophysical (e.g. geophysical location) and socio-economic (e.g. licensing contracts).

A similar co-construction of markets and natures is evident in the management and organization of *feedstock supply*. Access to biomass is only one aspect of the overall value chain, with the identification of a suitable feedstock being another critical element; for example, softwood trees are more suitable than hardwoods for conversion into biofuels, but both types of species grow together meaning it is difficult to harvest and deliver homogeneous feedstock supply. Critically, advanced biofuels cannot be the prime timber user, commercially-speaking, as the cost of prime timber—at between C$125 and C$150 per bone dry metric ton—is simply not economically viable for biofuels that are meant to compete with petroleum. Instead, feedstock supply for advanced biofuels production is only viable if it uses 'residues' from primary timber production; for example, sawdust, offcuts, leftovers, etc. These residues have both a materiality (e.g. residual biomass from other uses) and a socio-economic quality (e.g. framed as a costless natural resource) to them. Again, this means that advanced biofuels production is only viable where markets are currently limited, especially for 'residues' since valuing those residues (i.e. pricing them) would immediately make the bioenergy derived from them uncompetitive with petroleum.

A final example of this co-construction of markets and natures is evident in the *technology conversion* processes deployed to produce advanced biofuels. Residues represent the key resource for these processes because they are cheap, while the technological processes are expensive—this contrasts with conventional biofuels where technology is cheap but feedstock expensive (Calvert et al. 2017b). Consequently, it is necessary to make the technology conversion processes 'feedstock agnostic' so that they can convert all sorts of 'residual' biomass into bioenergy. A critical reason for this is that the feedstock residues—discussed above—are diverse and because the biomass harvested is not homogeneous, neither in terms of tree species (with 6–8 main species across

Ontario) nor timber grades (ranging from knotty to sawdust). Moreover, the technology conversion processes have to produce a homogeneous, or fungible, commodity (i.e. sugar) from a heterogeneous feedstock (i.e. timber). However, creating a fungible commodity—meaning it does not matter who produces it since its quality is the same whoever does so—is dependent on the socio-material configuration of production, in that fungibility results from the infrastructure put in place to get a product to market rather than from qualities inherent within the product.

3.4 Conclusion

In outlining the co-construction of markets and natures in the bioeconomy, it is clear that there is more going on here than the insertion of markets into an otherwise pristine or untouched nature. It is important to stress that I am not trying to say that the neoliberal natures literature is necessarily wrong. Rather, I am trying to emphasize that markets are instituted through and within nature; they are not aberrations of them: a market can only be instituted through the co-construction of biophysical materialities and socio-economic configurations. I thereby emphasize the inherent contingency of this process and highlight that we can actively identify points in this *socio-material* instituting (cf. Polanyi 2001 [1944]) at which we may want to intervene to shift or transform the process itself or its outcomes. As such, we can choose the bioeconomies that we want to see emerge (Kitchen and Marsden 2011).

The political implications of this are that we need to understand how markets and natures are produced together, rather than one being an imposition on or aberration of the other. When it comes to the bioeconomy, for example, this approach provides the means to unpack the manner in which policy tools and biophysical materialities configure bioeconomies in certain ways, opening up room to intervene in the process. In the context of Canada, and especially Ontario, this is evident in the way that the Provincial Government enters into specific understandings and socio-material arrangements that configure forests as a 'resource' (Bridge 2009). Forests are made into resources through the Provincial Government's claim to ownership of 'Crown Land', their management of long-term harvesting licensing agreements, publicly funded access roads and support, and so on. This is not a recent phenomenon, nor is it a quick release of 'natural' assets (Birch and Muniesa 2020), but rather it is a reflection of a long-term and ongoing process (see Wang 2019 on edamame production for a similar example). Making alternative bioeconomies would entail picking apart the social *and* material arrangements in this current configuration, which might include handing forest lands back to indigenous First Nation bands, rethinking forest management or an end to logging roads.

Acknowledgements I acknowledge the financial support of the Social Sciences and Humanities Research Council of Canada for funding this research (Ref.: 430-2013-000751).

References

Amadae, S.A. (2016). *Prisoners of Reason.* Cambridge: Cambridge University Press.

Bakker, K. (2009). Commentary: Neoliberal Nature, Ecological Fixes, and the Pitfalls of Comparative Research. *Environment and Planning A, 41,* 1781–1787.

Bakker, K. (2010). The Limits of 'Neoliberal Natures': Debating Green Neoliberalism. *Progress in Human Geography, 34*(6), 715–735.

Becker, S., Moss, T., & Naumann, M. (2016). The Importance of Space: Towards a Socio-Material and Political Geography of Energy Transitions. In L. Gailing & T. Moss (Eds.), *Conceptualizing Germany's Energy Transition* (pp. 93–108). London: Palgrave Macmillan.

Bigger, P., & Dempsey, J. (2018). The Ins and Outs of Neoliberal Natures. *Environment and Planning E: Nature and Space, 1*(1–2), 1–51.

Birch, K. (2015). *We Have Never Been Neoliberal: A Manifesto for a Doomed Youth.* Winchester: Zero Books.

Birch, K. (2016a). Emergent Policy Imaginaries and Fragmented Policy Frameworks in the Canadian Bio-Economy. *Sustainability, 8*(10), 1–16.

Birch, K. (2016b). Materiality and Sustainability Transitions: Integrating Climate Change into Transport Infrastructure in Ontario, Canada. *Prometheus: Critical Studies in Innovation, 34*(3–4), 191–206.

Birch, K. (2017). *A Research Agenda for Neoliberalism*. Cheltenham: Edward Elgar.

Birch, K. (2019). *Neoliberal Bio-Economies? The Co-construction of Markets and Natures*. London: Palgrave Macmillan.

Birch, K., & Calvert, K. (2015). Rethinking 'Drop-In' Biofuels: On the Political Materialities of Bioenergy. *Science and Technology Studies, 28*, 52–72.

Birch, K., & Muniesa, F. (Eds.) (2020). *Assetization: Turning Things into Assets in Technoscientific Capitalism*. Cambridge: The MIT Press.

Birch, K., Levidow, L., & Papaioannou, T. (2010). Sustainable Capital? The Neoliberalization of Nature and Knowledge in the European Knowledge-Based Bio-Economy. *Sustainability, 2*(9).

Birner, R. (2018). Bioeconomy Concepts. In I. Lewandowski (Ed.), *Bioeconomy* (pp. 17–38). Cham: Springer.

Bridge, G. (2009). Material Worlds: Natural Resources, Resource Geography and the Material Economy. *Geography Compass, 3*(3), 1217–1244.

Bugge, M., Hansen, T., & Klitkou, A. (2016). What Is the Bioeconomy? A Review of the Literature. *Sustainability, 8*, 1–22.

Calvert, K., Birch, K., & Mabee, W. (2017a). New Perspectives on an Old Energy Resource: Biomass and Emerging Bio-Economies. In S. Bouzarovski, M. Pasqualetti, & V. Castan Broto (Eds.), *The Routledge Research Companion to Energy Geographies* (pp. 47–60). London: Routledge.

Calvert, K., Kedron, P., Baka, J., & Birch, K. (2017b). Geographical perspectives on sociotechnical transitions and emerging bio-economies: introduction to a special issue. *Technology Analysis & Strategic Management, 29*(5), 477–485.

Castree, N. (2008a). Neoliberalising Nature: The Logics of Deregulation and Reregulation. *Environment and Planning A, 40*, 131–152.

Castree, N. (2008b). Neoliberalising Nature: Processes, Effects, and Evaluations. *Environment and Planning A, 40*, 153–173.

Castree, N. (2010a). Neoliberalism and the Biophysical Environment 1: What 'Neoliberalism' Is, and What Difference Nature Makes to It. *Geography Compass, 4*(12), 1725–1733.

Castree, N. (2010b). Neoliberalism and the Biophysical Environment 2: Theorising the Neoliberalisation of Nature. *Geography Compass, 4*(12), 1734–1746.

Collard, R-C., Dempsey, J., & Rowe, J. (2016). Re-regulating Socioecologies Under Neoliberalism. In S. Springer, K. Birch & J. MacLeavy (Eds.), *The Handbook of Neoliberalism* (pp. 455–465). New York: Routledge.

Daemmrich, A. (2015). Anticipatory Markets: Technical Standards as a Governance Tool in the Development of Biodegradable Plastics. In S. Borras & J. Edler (Eds.), *The Governance of Socio-Technical Systems* (pp. 49–69). Cheltenham: Edward Elgar.

Dardot, P., & Laval, C. (2014). *The New Way of the World*. London: Verso.

Dempsey, J., & Robertson, M. (2012). Ecosystem Services: Tensions, Impurities, and Points of Engagement within Neoliberalism. *Progress in Human Geography, 36*(6), 758–779.

El-Chickakli, B., von Braun, J., Lang, C., Barben, D., & Philp, J. (2016). Five Cornerstones of a Global Bioeconomy. *Nature, 535*, 221–223.

Essex, J. (2016). The Neoliberalization of Agriculture: Regimes, Resistance, and Resilience. In S. Springer, K. Birch & J. MacLeavy (Eds.), *The Handbook of Neoliberalism* (pp. 514–525). New York: Routledge.

European Commission (2012). *Innovation for Sustainable Growth: A Bioeconomy for Europe: Communication from the Commission to the European Parliament, the Council, The European Economic and Social Committee and the Committee of the Regions*. Brussels. http://ec.europa.eu/research/bioeconomy/pdf/official-strategy_en.pdf. Accessed 20 Nov 2020.

Fletcher, R. (2014). Taking the Chocolate Laxative: Why Neoliberal Conservation "Fails Forward". In B. Büscher, W. Dressler, & R. Fletcher (Eds.), *Nature™ Inc: Environmental Conservation in the Neoliberal Age* (pp. 87–107). Tucson: University of Arizona Press.

Foucault, M. (2008). *The Birth of Biopolitics*. New York: Picador.

Friedman, M. (1962). *Capitalism and Freedom*. Chicago: University of Chicago Press.

Frow, E., Ingram, D., Powell, W., Steer, D., Vogel, J., & Yearley, S. (2009). The Politics of Plants. *Food Security, 1*(1), 17–23.

German Bioeconomy Council (2015a). *Bioeconomy Policy: Synopsis and Analysis of Strategies in the G7*. Berlin: Office of the Bioeconomy Council. https://biooekonomierat.de/fileadmin/international/Bioeconomy-Policy_Part-I.pdf. Accessed 23 Nov 2020.

German Bioeconomy Council (2015b). *Bioeconomy Policy (Part II): Synopsis of National Strategies Around the World*. Berlin: Office of the Bioeconomy Council. https://biooekonomierat.de/fileadmin/international/Bioeconomy-Policy_Part-II.pdf. Accessed 23 Nov 2020.

Harvey, D. (2005). *A Brief History of Neoliberalism*. Oxford: Oxford University Press.

Hausknost, D., Schriefl, E., Lauk, C., & Kalt, G. (2017). A Transition to Which Bioeconomy? An Exploration of Diverging Techno-Political Choices. *Sustainability, 9*(4), 1–22.

Hayek, F. (2001 [1944]). *The Road to Serfdom*. London: Routledge.

Hayek, F. (2011 [1960]). *The Constitution of Liberty*. Chicago: Chicago University Press.

Kenney-Lazar, M., & Kay, K. (2017). Value in Capitalist Natures. *Capitalism Nature Socialism, 28*(1), 33–38.

Kitchen, L., & Marsden, T. (2011). Constructing Sustainable Communities: A Theoretical Exploration of the Bio-Economy and Eco-Economy Paradigms. *Local Environment, 16*, 753–769.

Larner, W. (2003). Neoliberalism? *Environment and Planning D, 21*, 509–512.

Levidow, L., Birch, K., & Papaioannou, T. (2012). EU Agri-Innovation Policy: Two Contending Visions of the Knowledge-Based Bio-Economy. *Critical Policy Studies, 6*, 40–66.

Loftus, A., & Budds, J. (2016). Neoliberalizing water. In S. Springer, K. Birch & J. McLeavy (Eds.), *The Handbook of Neoliberalism* (pp. 503–513). New York: Routledge.

Lohmann, L. (2016). Neoliberalism's Climate. In Springer, S., Birch, K., & MacLeavy, J. (Eds.), *The Handbook of Neoliberalism* (pp. 480–492). London: Routledge.

McAfee, K. (2003). Neoliberalism on the Molecular Scale: Economy and Genetic Reductionism in Biotechnology Battles. *Geoforum, 34*, 203–219.

McCarthy, J., & Prudham, S. (2004). Neoliberal Nature and the Nature of Neoliberalism. *Geoforum, 35*, 275–283.

McKibben, B. (2012). Global Warming's Terrifying New Math. *Rolling Stone*, 19 July. https://www.rollingstone.com/politics/news/global-warmings-terrifying-new-math-20120719. Accessed 7 Oct 2020.

McCormick, K., & Kautto, N. (2013). The Bioeconomy in Europe: An Overview. *Sustainability, 5*, 2589–2608.

Mirowski, P. (2013). *Never Let a Serious Crisis Go to Waste*. Cambridge: Harvard University Press.

Moore, J.W. (2015). *Capitalism in the Web of Life*. London: Verso.

Mukhtarov, F., Gerlak, A., & Pierce, R. (2017). Away from Fossil-Fuels and Toward a Bioeconomy: Knowledge Versatility for Public Policy? *Environment and Planning C, 35*(6), 1010–1028.

Peck, J., & Tickell, A. (2002). Neoliberalizing Space. *Antipode, 34*(3), 380–404.

Pfau, S.F., Hagens, J.E., Dankbaar, B., & Smits, A.J.M. (2014). Visions of Sustainability in Bioeconomy Research. *Sustainability, 6*, 1222–1249.

Polanyi, K. (2001 [1944]). *The Great Transformation*. New York: Beacon Press.

Prudham, S. (2005). *Knock on Wood: Nature as Commodity in Douglas-Fir Country*. New York: Routledge.

Richardson, B. (2012). From a Fossil-Fuel to a Biobased Economy: The Politics of Industrial Biotechnology. *Environment and Planning C, 30*, 282–296.

Roff, R.J. (2008). Preempting to Nothing: Neoliberalism and the Fight to De/Re-regulate Agricultural Biotechnology. *Geoforum, 39*, 1423–1438.

Schmid, O., Padel, S., & Levidow, L. (2012). The Bio-Economy Concept and Knowledge Base in a Public Goods and Farmer Perspective. *Bio-based and Applied Economics, 1*, 47–63.

Staffas, L., Gustavsson, M., & McCormick, K. (2013). Strategies and Policies for the Bioeconomy and Bio-Based Economy: An Analysis of Official National Approaches. *Sustainability, 5*, 2751–2769.

Springer, S., Birch, K., & MacLeavy, J. (Eds.) (2016). *The Handbook of Neoliberalism*. New York: Routledge.

The White House (2012). *National Bioeconomy Blueprint*. Washington, DC. https://obamawhitehouse.archives.gov/sites/default/files/microsites/ostp/national_bioeconomy_blueprint_april_2012.pdf. Accessed 23 Nov 2020.

Tyfield, D. (2017). *Liberalism 2.0 and the Rise of China*. London: Routledge.

Wang, K-C. (2019). The Art of Rent: The Making of Edamame Monopoly Rents in East Asia. *Environment and Planning E*.

Open Access This chapter is licensed under the terms of the Creative Commons Attribution 4.0 International License (http://creativecommons.org/licenses/by/4.0/), which permits use, sharing, adaptation, distribution and reproduction in any medium or format, as long as you give appropriate credit to the original author(s) and the source, provide a link to the Creative Commons license and indicate if changes were made.

The images or other third party material in this chapter are included in the chapter's Creative Commons license, unless indicated otherwise in a credit line to the material. If material is not included in the chapter's Creative Commons license and your intended use is not permitted by statutory regulation or exceeds the permitted use, you will need to obtain permission directly from the copyright holder.

4

Tools of Extraction or Means of Speculation? Making Sense of Patents in the Bioeconomy

Veit Braun

4.1 Introduction

What is the source of value in the bioeconomy? The conclusion that might be drawn from the various national and international bioeconomy strategies (Birch and Tyfield 2012; Backhouse et al. 2017)—that organisms, ecosystems and biological processes are infinite sources of energy, industrial raw materials and foodstuffs—is one that is challenged by the contributions to this volume. As many of the pieces in this volume show, the goods associated with the bioeconomy do not come from the green land of plenty: more often than not, they are produced under dire working conditions, at the expense of existing social and economic structures and with ecological consequences that complicate the idea of the bioeconomy as a sustainable, non-extractive way of life. But what about the fruits of the mind? After all, the bioeconomy is not just a

V. Braun (✉)
Goethe University Frankfurt, Frankfurt am Main, Germany
e-mail: braun@soz.uni-frankfurt.de

vision of agro-industrial production. In the "knowledge-based bioeconomy" (Birch and Tyfield 2012), the basis for growth and value is not so much life's ability to produce surplus (Cooper 2008) but the value added by technological innovation. This does not simply mean that science and technology are what makes nature or life productive in the first place. As a continuation of the older idea of the "knowledge economy", it might even suggest that knowledge about biological objects and processes can itself be turned into a good (Slaughter and Rhoades 2004). In Birch's and Tyfield's (Birch and Tyfield 2012, p. 308) words,

> What this would imply is that it is the knowledge and knowledge labour required to transform these fragments into commodities that are valuable, implying that the prefix 'bio-' is rather irrelevant in this case. We may as well term it 'knowledge-'value instead since it is knowledge (or, more accurately, knowledge labour) which creates value and not the latent qualities of the biological material itself.

What, then, does value consist of and where does it flow from in the bioeconomy? In this chapter, I want to tackle this question by looking at the empirical case of European patents on plants.[1] Since the conception of the knowledge-based economy, especially in the life sciences, patents have provided a crucial link between science and industry, providing the legal basis for managing innovations as a private rather than a public good (Slaughter and Rhoades 2004). The "enclosure" of knowledge in patents, the economic argument goes, helps to turn elusive intangible knowledge into something resembling classical tangible goods (Landes and Posner 2003). While from a business point of view this is necessary to fend off competitors and to recover R&D investments, the established critical argument against patents is that they deprive the public of an otherwise accessible good—unnecessarily so, it is pointed out, because knowledge, unlike tangible goods, is neither scarce nor exhaustible. The counterargument consists in pointing to positive externalities generated

---

[1]This chapter is based on my PhD thesis *Seed at the End of Property: Propertization in Plant Breeding and its Crises* (LMU Munich, 2018). Methods included participant observation, interviews with breeders, lobbyists, patent lawyers and managers as well as document analysis and patent statistics.

through the disclosure of knowledge and the public availability of new technologies after the expiration of patents. The debate is thus about the tension between the positive value of patents to firms and their negative value to the public (cf. Parthasarathy 2017), and whether the latter is justified or compensated for by public benefits further down the line (Bently and Sherman 2014, pp. 379–381).

Does this long-standing controversy adequately capture the functions and effects of patents in the bioeconomy? Both the affirmative and the critical economic theory of patents assume a transsubstantiation of knowledge into tangible products, with the latter constituting the "real" value of the knowledge enclosed, as they can be turned into money on commodity markets. This idea has been challenged by patent research. Only about 5 to 10% of all granted patents result in a commercial product, with the number of commercially *successful* patents being even lower (Schankerman and Pakes 1986). Birch (2017) even argues that there are virtually no commercial products in the bioeconomy that would explain the level of market capitalisation. Does this mean that there is no value to patents in the bioeconomy at all? If so, then both the classical arguments for and against patents would be moot. If not, then how are we to understand the value of patents and the nature of what is protected by them?

As I will show in the following, the answer to the question of patents and value in the bioeconomy is not straightforward: plant patents are used in various ways, many of them in a manner that is counterintuitive to the established economic theory of patents. European plant patents on conventional seed complicate critical stances towards patents and their role in "knowledge-based" economies as well as the notions of invention and knowledge. Patents are pursued for very different, often contradictory reasons—protecting sales and firms, fending off competitors, facilitating cooperation or signalling value. Markedly different from biotech patents, patents on non-GM plants highlight the complexity of valuing nature while foreshadowing the conflicts to come over gene-edited plants. This also poses a challenge for a generalised critique of patenting in the bioeconomy.

4.2 From Biotech to Native Traits

The past 30 years have seen an increase in patent applications for conventionally bred plants at the European Patent Office (EPO), with a steep rise in applications from the late 1990s onward. This trend is a result of a coalescence of various scientific, economic and legal developments. Although breeding native traits has been possible for at least a century—if not during the millennia since the agricultural revolution—patents on such traits are a relative novelty. The simultaneous rise of conventional and biotech patents in the 1990s suggests that native trait patents follow the model of patents on genetically modified plants, which were first developed and patented in the mid-1980s (Charles 2001).

This technological breakthrough, together with understanding genes as chemical compounds (Calvert and Joly 2011), leads to a paradigm shift in plant breeding. Plants, long deemed unpatentable (Pottage and Sherman 2011), could now be treated as inventions. Patent law was reinterpreted and reformed across the globe in the 1980s and 1990s to facilitate the patenting of transgenic plants, which in 1998 resulted in the EU Biotechnology Directive (Parthasarathy 2017). However, GM food was widely opposed by European consumers, supermarkets and farmers, preventing a green biotech industry from emerging. Today, plants are almost exclusively bred by conventional techniques—crossing, selecting and back-breeding—occasionally supplemented by laboratory techniques like cell culture and double haploids. The firms in the European seed sector include large plant science multinationals, but also a considerable number of medium- or moderately large-sized seed producers that often specialise in a small number of crop species (Ragonnaud 2013; Brandl 2018).

Although not as costly as genetic transformation, which costs about USD 100 million (Phillips McDougall 2011), breeding new plant traits with conventional means still takes around 10 to 15 years and between 1 and 2 million euros (Goodman 2002). The plant variety protection (PVP) law provides an opportunity to protect intellectual property right related to plants in Europe, but it only applies to finished varieties, not to new individual plant traits (Sanderson 2017; Braun 2020). PVP gives breeders a temporary monopoly on their varieties, but allows third parties to use these varieties as sources of traits or to develop new varieties. Firms can therefore recoup investments in variety development, whereas trait breeding is economically less attractive. For this reason, trait development in Europe is usually delegated to public research institutes or organised collectively among private breeders with some public assistance. In the latter case, a number of companies will share the costs of developing a new trait while coordinating market introduction of their varieties (Becker 2011; Brandl 2018). Overall, European breeders support these public and collaborative models for trait development and consider them sufficient and effective.

4.3 Patenting Native Traits: Shifts in the Legal Landscape in Europe

The 1990s legal reforms opened a backdoor for patents not just on biotech but also on conventional plants. In March 2015, the EPO's Enlarged Board of Appeal set out its ruling in the "Tomato/Broccoli II" case. The ruling stated that while conventional varieties and breeding techniques themselves could not be patented due to their exemption under European patent law, the products of such techniques—conventionally bred traits—could. The reasoning behind the decision was that there was no explicit exclusion of traits; biotechnology traits had, in fact, explicitly been declared patentable. If the latter were compatible with PVP, there was no reason to assume that plant traits bred with conventional means were not. These are commonly called "native traits" because they originate in the crop species or genus itself rather than from a genetically very different organism (Girard 2015; Kock 2017). As such, they can be transferred with the established techniques of deliberate crossing, selecting and back-breeding, which have been at the heart of modern plant breeding since the nineteenth century. In the wake of green biotechnology, however, these native traits also attracted the interest of patent lawyers. Between 1981 and 2018, 438 patent applications were made for native traits, 117 of which were granted (Patstat 2019). While these figures may not appear to be very high, they are in about the same range as those for biotech plants.

If trait development is by and large economically unattractive in Europe, what is attractive about patents of the tomato/broccoli type? Why have these patents become controversial, especially for the seed industry? Unlike plant biotechnology, which is more or less restricted to four species of field crops (soy, maize, cotton and rapeseed), the range of species in native trait patents is more diverse and surprisingly vegetable-heavy. Out of 117 granted patents, between 33 and 37 (depending on the way in which "vegetable" is defined) fall into this category.[2] Furthermore, specialised vegetable breeders are well represented among applicants. Although largely absent during the 1990s (with just 8.4% of all applications up to 1999), they accounted for 21.4% of applications filed between 2000 and 2017. The major patent disputes in the field of native traits have so far been about vegetables. Meanwhile, Monsanto, Syngenta and Bayer have all strongly invested in vegetable breeding programmes and the vegetable sector has seen a strong market concentration (Ragonnaud 2013), which is reminiscent of the North American market in the wake of agricultural biotech (Schenkelaars et al. 2011).

There is a slight trend towards "output traits" in the analysed patents: while input traits, which dominate plant biotechnology, primarily benefit farmers, output traits like enhanced nutritional content (like the broccoli patent) or facilitated processing (as in the tomato patent) target processors and consumers in the first place. Vegetables have higher profit margins than field crops; both seed and fruit production in greenhouses are more labour-intensive than field agriculture (Becker 2011), meaning that R&D investments make up a smaller share of the total costs and are thus more easily recovered. There is, however, also a considerable number of pest resistance traits among the sample; native traits patents, thus, are not simply "output trait" patents.

---

[2]The classification of *Brassica* species in patents is not straightforward, as they often refer to the whole genus or to field (e.g. rapeseed) as well as greenhouse crops (e.g. broccoli).

4.4 Tools of Extraction?

There is some evidence that patented traits mirror the unequal, often extractive economic relations between the Global North and the Global South highlighted by other contributions to this volume. Many of these traits originate in so-called exotic materials or wild relatives of crop species (Acquaah 2012). While the discussion is mainly focused on innovations and inventions that occur in Europe, in other words, the seed companies' nurseries, the genetic foundation of pest resistance, self-drying and other plant abilities often comes from the centres of diversity, overwhelmingly located in the Global South (Kloppenburg 2004). The tomato patent, for example, makes claims on traits derived from a wild tomato relative found in Peru and Ecuador. The broccoli patent derives its glucosinolates from a threatened Sicilian species of cabbage, while a patent by Syngenta on bell pepper, also subject to extensive litigation at the EPO, is based on Jamaican wild pepper germplasm (Leberecht and Meienberg 2014).

The flow and subsequent valuation of genes, commonly labelled "biopiracy", from the developing south to the industrial north has been a long-standing topic in postcolonial discourse (Hayden 2003) and has a tradition reaching back to the beginnings of colonial ethnobotany (Schiebinger 2011; Brockway 2011). The process, however, is usually not as straightforward as a company venturing into the tropics to screen and collect plants and ship them back home for breeding. More often, seed banks serve as intermediaries by collecting, describing and storing plant germplasm. Seed companies then access this pre-collected material rather than original populations in the Global South. Syngenta's pepper is a well-documented example; the original material is from a publicly funded expedition in the 1970s, predating private bioprospecting (Leberecht and Meienberg 2014). Nevertheless, the contribution of countries of origin and their native communities remains a contested topic (Bertacchini 2008).

Another dominant narrative in the critical scholarship on agricultural biotechnology and the study of intellectual property has been the rise of "neofeudal" property relations engendered by patents and other intellectual property (IP) rights (Schubert et al. 2011; Perzanowski and

Schultz 2018; Braun 2020). In these asymmetrical relations, IP rights are not used as horizontal instruments to fend off competitors but as vertical tools to control property objects and, by extension, processors and farmers beyond the point of sale (Kloppenburg 2014). Lending legal existence to and conferring ownership of technologies, patents give companies a means to license rather than sell their seed. On top of creating new business models, which only work if seed companies retain control over the seed, licensing also enables companies to evade competition and force additional terms upon their licensees (Perzanowski and Schultz 2018), extracting additional profits from actors not upstream, but downstream in the value chain. In the case of biotech seed, this involves the restriction that licensed seed may only be used in combination with chemical products from the same company, forcing farmers to buy the whole package, even if they only want the seed (Schubert et al. 2011).

While patents permit the licensing of vegetable seeds and most vegetable breeders are vertically integrated (i.e. breeding, seed production and distribution take place in one company), it is unlikely that the rise in vegetable patents is driven by such business models. Licensing seed to farmers is already possible with PVP and, in fact, widespread in tomato production (tomato grower, Interview no. 1, August 2017). Thus, patents do not add an additional instrument with which to capture a higher share of the profits across the value chain. Few native traits have been a commercial success: even though patent disputes are generally regarded as indicative of high commercial value in the industry (former head of IP, Interview no. 2, June 2018), the tomato patent, for example, has never been used for breeding actual self-drying tomatoes (patent lawyer, Interview no. 3, November 2017). Indeed, only 25 patented traits were present in commercial varieties in 2016 (Kock and ten Have 2016). Overall, native trait patents in Europe, unlike their biotech counterparts in North America, have not led to new vertical property relations between breeders and farmers.

4.5 Using by not Using: Traditional Breeders and Native Trait Patents

If the value of native trait patents does not rest in superior products or the ability to squeeze additional profits from actors further down the value chain, where else could it lie? Studies in business and management have identified a number of "noncommercial" uses for patents, i.e. patents applied for and granted but never turned into products (Torrisi et al. 2016). These follow different logics: a firm might not "use" a patent, as its value rests in blocking a competitor from access to a specific technology. Such patents are often redundant to other patents the firm owns, preventing rival companies from "inventing around" them (Landes and Posner 2003, p. 295). Alternatively, patents can be used as "bargaining chips" (Noel and Schankerman 2013), giving companies leverage in merger and acquisition negotiations or licensing agreements. An even more profane reason for patenting technologies is not so much to prevent others from accessing them but the reverse, i.e. preventing a lock-out from central technologies (Torrisi et al. 2016).

In one form or another, these strategies can also be found among the holders of native trait patents. This group is notably diverse: it includes public research institutes such as INRA (France), CSIC (Spain) or CSIRO (Australia); plant science multinationals and their vegetable subsidiaries like Nunhems (formerly Bayer, now BASF), Seminis and De Ruiter (both formerly Monsanto, now Bayer); and independent, mostly Dutch seed producers such as Bejo, Enza or Rijk Zwaan. While the public applicants have pursued technology diffusion strategies with their patents as part of new public management (Jewell 2015), the motives of private applicants are more varied. Some of the latter have opposed native trait patents, either specific ones or in general. One example here is Syngenta (32 applications), which, otherwise pro-patent, contested the broccoli patent; the independent Dutch companies that take an explicit anti-patent stance yet engage in patenting, sometimes holding extensive trait portfolios, are further examples.

In an interview, representatives of one firm made it clear that patenting their traits was a defensive strategy:

We do it. […] But the reason for that is that we saw this development by technology, more and more patents. And, also, the consolidation; bigger and bigger companies were created. Because in the 1990s we had Syngenta, there was SaatUnion still [an] independent company, and they are not completely independent anymore. We had Royal Sluis, which later, through a lot of takeovers, became part of Monsanto, Seminis-Monsanto. We thought, okay, we do not need a patent protection for the [plants], because we have plant breeders' right[s] for that. But we need the patents to protect ourselves. Because if all these big multinationals – and [our firm] is, compared to the multinationals, still small, but at that time it was really small. […] For example, […] if a competitor would have a patent on a very important characteristic in lettuce, and we couldn't get a license, then it would mean that we miss half of our turnover. And that is the end of the company. (head of IP, Interview no. 4, December 2017)

The company has not developed a business model around its patents and has no plans to do so. Patented traits are licensed to competitors for a small fee, which barely covers the costs of applying for and renewing the patents: due to economies of scale in R&D, assessment and litigation, it is more expensive for a small firm to engage in patenting than for a big one (Lanjouw and Schankerman 2001). For this company, the patent portfolio serves as insurance against hostile competitors who might use their own patents to squeeze it out of the market.

At the same time, the company's patent portfolio has forced bigger firms to take it seriously in negotiations and strike cross-licensing deals. While not all independent breeders with trait patents have made similar deals, their attitudes to patenting are likely similar. Traditional breeding companies argue that trait development can sufficiently be covered with revenues from variety development and sale; they do not see a need for patents on traits or a market for trait innovations. Instead, they stress the negative impact of patents on the freedom to breed under PVP, high transaction costs attached to patents and legal and economic insecurity tied to working with patented material.

4.6 Speculation, Not Innovation? Patents as Credit and Capital

But what about the plant science multinationals? Is their approach to trait patents really an exclusively aggressive one? As we have seen, most patents in the field never find commercial use. This does not necessarily mean that biotech companies would not resort to such strategies if they had a patent on a major native trait. In fact, the history of mergers and acquisitions in the 1990s and 2000s in the plant breeding industry should make conventional breeders wary. There are, however, other explanations for why plant science firms invest time, money and other resources in patenting in a comparatively small market without notable commercial success. The various non-commercial strategies pursued by smaller competitors obviously also apply to bigger firms: if the former need patent portfolios for defence and as bargaining chips, so do multinationals.

Multinationals have additional motives for pursuing patents, though. As Kang (2020) shows, patents have become a value of their own in many publicly traded companies. Uncoupled from the total sum of the sales they protect, they primarily serve as a signal to investors and rating firms. In this logic, a patenting firm is a firm with an active and successful R&D department, a firm that produces value—in short, an *innovative* firm. At the same time, however, as patents are valued as such and not with reference to products or sales, they become empty signifiers proclaiming but not actually containing value. Kang's analysis echoes a wider criticism of the bioeconomy and life science research as "a passel of Ponzi schemes" (Mirowski 2012). Like in the late 1990's Dotcom economy, this critique points out that there are no "real" values to back up the capitalisation of companies on the stock markets (Birch 2017).

A patent on a native trait thus need not be commercially important as long as it adds to the company's patent portfolio, signalling to shareholders and investors that their money is well invested in an active company with a steady output of innovations. Patenting a trait black-boxes the question of its actual value (Kang 2020) in the legal-economic object of the patent. This is why life science start-ups put all their efforts into obtaining a patent (Haeussler et al. 2014): it is what constitutes their value in the eyes of a potential buyer. Indeed, actors in the seed business confirm that such a "post-commodity" mindset (cf. Birch and Muniesa 2020) is widespread among plant science multinationals, as in this statement from a breeder working for a subsidiary of a biotech company:

> Someone [at headquarters] has an idea for a project: 'We're making hybrid wheat now!' So the person goes to their superior and says: 'Boss, let's make hybrid wheat. With hybrid wheat, we'll […] get 30% of the global wheat market, which amounts to 20 billion euros. All I need is 20 million euros and 50 people.' And the boss […] tells this to the board of directors, who have even less of a clue about wheat breeding. But all they hear is 30% of the market for 20 million. So they approve it. Then the firm issues a press release […] The next thing that happens is that stocks go up because of that release. So [my firm] goes to its shareholders and says 'Look, you just made 25 euros profit on every single share you have thanks to our great business performance. Out of those 25 euros, could you maybe give us 5 and keep the other 20 so that we can reinvest it to make you even more money?' And then the whole thing starts all over again. (plant breeder, Interview no. 5, May 2017, paraphrased)

The fetishism of projects, innovations and patents is not just a scheme for tricking investors. It is mirrored in internal communications, decisions and rewards within the company. Patents serve as a currency between superiors and subordinates as well as between a firm's departments: there is a tendency for self-reinforcing feedback loops to develop in their interaction. R&D departments produce patentable innovations to justify hires and expenditures. IP departments turn these innovations into patents to legitimise their existence, controlling lists of patents among the company's performance indicators because they are easily quantified and labelled as valuable assets (Long 2002; Hsu and Ziedonis 2008; Gill et al. 2012).

Public companies' patent portfolios in native traits could thus be explained as aimed at the stock rather than the seed markets. The lack of commercial value would not contradict their function as internal and external signal of innovativeness and company value. And even if smaller firms use patents differently in their interactions with other market actors, this need not prevent similar logics from unfolding inside the firm: the self-referential nature of "patents as credit" (Kang 2015) largely emerges from complexity within companies of a certain size.

However, plant science multinationals do not necessarily agree with this theory. This is not necessarily because they need to uphold (both their own and others') belief in patents as meaningful indicators of innovativeness, but because some of them have diagnosed the pitfalls of a self-referential patent economy themselves and taken countermeasures:

> My KPIs [key performance indicators] were cost management. So basically, we made sure that the patents we have are actually in active use. Either in relation to products or not [in use]. My KPI is efficiency-driven. Reducing costs, keeping only what is actually relevant […]. So in number of patents – yes, I know, [other firms have] KPIs like 'We need to make at least 1000 patents a year,' but that's complete bullshit. And no educated investor will buy that anymore either. So it's quality, not quantity. If I can protect my portfolio of products with a smaller number of patents, then that's a considerable efficiency factor. The patent itself does not have any added value. The added value is only what's protected by it. And if it doesn't protect, if I enclose a piece of desert, then I'm wasting money! So [that's] a very outdated vision; it's done much more pragmatically nowadays. (former head of IP, Interview no. 2, June 2018)

This quote confirms the problem of self-referentiality but also situates it in the past: under his tenure, the interviewee points out, his company took a radically different approach to its patent portfolio. Unlike other companies, which simply counted their patents, he linked patents back to physical sales and only kept those that covered their costs. Recent publications in management literature argue for a similar turn away from portfolio size to quality and management of patents (Ernst 2017).

In addition, some firms in the business, notably Syngenta, have pushed for changes in European plant patent law while cutting their patent portfolios considerably. Syngenta spearheaded the International Licensing Platform for Vegetables (ILP Vegetable), a clearinghouse meant to drastically reduce the costs of licensing and breeding with patented traits (Kock and ten Have 2016; van Overwalle 2017). In exchange for access to patented traits, members have to make their patents available under the same conditions. Unlike in classical bilateral negotiations, licence fees cannot be set arbitrarily or prohibitively. If two parties cannot agree, they have to submit their bids to a committee, which then has to pick one over the other. The losing party will then have to cover the costs of the process.

ILP Vegetable's vision is to streamline patents into a useful tool for trait breeding. The platform is meant to reduce many of the transaction costs and uncertainties around patents, making them more accessible to smaller firms while at the same time preventing the use of patents as tools for aggressive monopolisation. However, this would also require conventional plant breeders to embrace patents as some form of market instrument, a commitment they are currently refusing to make. In addition, private solutions rely on voluntary participation, leaving open the possibility of aggressive patent use in the sector. Finally, there are also some legal uncertainties around the legal feasibility of an almost industry-wide pooling of patents (van Overwalle 2017; contra: Kock and ten Have 2016).

The platform nevertheless demonstrates that there are ideas and efforts directed at the proper use and actual value of patents in the industry that diverge from both the orthodox economic theory of patents and their various actual uses; these ideas and efforts should be taken seriously. Especially as the next wave of plant patents is just around the corner: with gene editing, many of the issues around native traits are about to return, albeit in a similar, yet different form. Gene-edited traits will be cheap and quick to produce, but also of a technical nature. With gene-edited traits proliferating in commercial varieties, breeders will face legal obstacles when trying to cross-breed with such plant material. Even if patent offices agree that native traits are "essentially biological" and, thus, unpatentable, this would not apply to gene-edited plants (Kock 2017). Once more, then, patent law would beat PVP law.

4.7 Conclusion: Patents in the Bioeconomy

If there is a lesson from the legal debate on native traits, it is first and foremost a negative one: we cannot definitely say what constitutes the value of patents in and for the bioeconomy. At least when it comes to patents, we cannot characterise the bioeconomy as predominantly based on extraction or speculation. Bio-patents are used in very different ways by different actors within the same industry. This defies any smooth policy narrative of patents as innovation incentives and vectors in the bioeconomy. However, it also complicates a critique of patents as instruments for extracting value or refeudalising economic relations. While refeudalisation captures an important and troubling element of the bioeconomy of GM seed, the case of native traits is different, despite shared genealogies and protagonists. The reason lies both in a specifically European situation (tensions between PVP and patent law, the makeup of the European plant breeding sector) and in the diverging biological nature of native traits (which are cheaper to produce than biotech traits and, thus, comprise different traits in different crop species). A similar difference will certainly manifest itself in gene-edited traits should there ever be considerable economic activity in that field in Europe.

However, there are also positive lessons to be drawn. While we have seen very different strategies and philosophies for patent use, they always referred and reacted to other strategies. Value is not "contained" in patents but unfolds in relation to sales, other patents, stock markets, competitors' perceptions and their won visions for the future of trait breeding. It is not an inherent quality of life itself, nor a simple effect of investment bubbles. Instead, whether patents are of value and if that value is positive or negative depends on one's situation in the complex landscape of European seed production. To determine the value of patents, we thus cannot simply look at them in isolation. Therefore, any policy that seeks to stimulate innovation in a bio-based economy needs to think beyond patents as a single-purpose instrument or the sole incentive for companies to innovate.

List of Interviews quoted

Interview no.	Profession/Function	Date and place
Interview no. 1	Tomato grower	08/2017, Germany
Interview no. 2	Former head of IP	06/2018, Switzerland
Interview no. 3	Patent lawyer	11/2017, Germany
Interview no. 4	Head of IP	12/2017, Netherlands
Interview no. 5	Plant breeder	05/2017, Germany

References

Acquaah, G. (2012). *Principles of Plant Genetics and Breeding* (2nd ed.). Hoboken: Wiley.

Backhouse, M., Lorenzen, K., Lühmann, M., Puder, J., Rodríguez, F., & Tittor, A. (2017). Bioökonomie-Strategien im Vergleich. Gemeinsamkeiten, Widersprüche und Leerstellen. *Working Paper Nr. 1, Bioeconomy & Inequalities*, Jena. https://www.bioinequalities.uni-jena.de/sozbemedia/neu/2017-09-28+workingpaper+1.pdf. Accessed 19 Oct 2020.

Becker, H. (2011). *Pflanzenzüchtung* (2nd ed.). Stuttgart: Verlag Eugen Ulmer.

Bently, L., & Sherman, B. (2014). *Intellectual Property Law* (4th ed.). Oxford: Oxford University Press.

Bertacchini, E.E. (2008). Coase, Pigou and the Potato: Whither Farmers' Rights? *Ecological Economics, 68*(1–2), 183–193.

Birch, K. (2017). Rethinking Value in the Bio-Economy: Finance, Assetization, and the Management of Value. *Science, Technology, & Human Values, 42*(3), 460–490.

Birch, K., & Muniesa, F. (2020). Introduction. In K. Birch & F. Muniesa (Eds.), *Turning Things into Assets* (pp. 1–43). Cambridge: MIT Press.

Birch, K., & Tyfield, D. (2012). Theorizing the Bioeconomy: Biovalue, Biocapital, Bioeconomics or … What? *Science, Technology & Human Values, 38*(3), 299–327.

Brandl, B. (2018). *Wissenschaft, Technologieentwicklung und die Spielarten des Kapitalismus: Analyse der Entwicklung von Saatgut in USA und Deutschland*. Wiesbaden: Springer VS.

Braun, V. (2020). From Commodity to Asset and Back Again: Value in the Capitalism of Varieties. In K. Birch & F. Muniesa (Eds.), *Turning Things into Assets* (pp. 203–224). Cambridge: MIT Press.
Brockway, L. (2011). Science and Colonial Expansion: The Role of the British Royal Botanic Gardens. In S.G. Harding (Ed.), *The Postcolonial Science and Technology Studies Reader* (pp. 110–126). Durham: Duke University Press.
Calvert, J., & Joly, J.-P. (2011). How Did the Gene Become a Chemical Compound? The Ontology of the Gene and the Patenting of DNA. *Social Science Information, 50*(2), 1–21.
Charles, D. (2001). *Lords of the Harvest: Biotech, Big Money, and the Future of Food*. Cambridge: Perseus Pub.
Cooper, M. (2008). *Life as Surplus: Biotechnology and Capitalism in the Neoliberal Era*. Seattle: University of Washington Press.
Ernst, H. (2017). Intellectual Property as a Management Discipline. *Technology & Innovation, 19*(2), 481–492.
Gill, B., Brandl, B., Böschen, S., & Schneider, M. (2012). Autorisierung. Eine wissenschafts- und wirtschaftssoziologische Perspektive auf geistiges Eigentum. *Berliner Journal für Soziologie, 22*(3), 407–440.
Girard, F. (2015). 'Though the Treasure of Nature's Germens Tumble All Together': The EPO and Patents on Native Traits or the Bewitching Powers of Ideologies. *Prometheus, 33*(1), 43–65.
Goodman, M.M. (2002). *New Sources of Germplasm: Lines, Transgenes, and Breeders*. Presented at the Memoria Nacional de Fitogenetica, Saltillo, Coah.
Haeussler, C., Harhoff, D., & Mueller, E. (2014). How Patenting Informs VC Investors—The Case of Biotechnology. *Research Policy, 43*(8), 1286–1298.
Hayden, C. (2003). *When Nature Goes Public: The Making and Unmaking of Bioprospecting in Mexico*. Princeton: Princeton University Press.
Hsu, D.H., & Ziedonis, R.H. (2008). Patents as Quality Signals for Entrepreneurial Ventures. *Academy of Management Proceedings, 2008*(1), 1–6.
Jewell, C. (2015, August). Who Benefits from IP Rights in Agricultural Innovation? *WIPO Magazine, 2015*(4). https://www.wipo.int/wipo_magazine/en/2015/04/article_0003.html. Accessed 19 Oct 2020.
Kang, H.Y. (2015). Patents as Credit: When Intellectual Property Becomes Speculative. *Radical Philosophy, 194*, 29–37.
Kang, H.Y. (2020). Patents as Assets: Intellectual Property Rights as Market Subjects and Objects. In K. Birch & F. Muniesa (Eds.), *Turning Things into Assets* (pp. 45–74). Cambridge: MIT Press.

Kloppenburg, J.R. (2004). *First the Seed: The Political Economy of Plant Biotechnology*. Madison: University of Wisconsin Press.

Kloppenburg, J.R. (2014). Re-purposing the Master's Tools: The Open Source Seed Initiative and the Struggle for Seed Sovereignty. *The Journal of Peasant Studies, 41*(6), 1225–1246.

Kock, M.A. (2017). Patenting Non Transgenic Plants in the EU. In D. Matthews & H. Zech (Eds.), *Research Handbook on Intellectual Property and the Life Sciences* (pp. 132–159). Cheltenham: Edward Elgar.

Kock, M.A., & ten Have, F. (2016). The 'International Licensing Platform—Vegetables': A Prototype of a Patent Clearing House in the Life Science Industry. *Journal of Intellectual Property Law & Practice, 11*(7), 496–515.

Pottage, A., & Sherman, B. (2011). Kinds, Clones, and Manufactures. In M. Biagioli, P. Jaszi, & M. Woodmansee (Eds.), *Making and Unmaking Intellectual Property* (pp. 269–283). Chicago: University of Chicago Press.

Landes, W.M., & Posner, R.A. (2003). *The Economic Structure of Intellectual Property Law*. Cambridge: Harvard University Press.

Lanjouw, J.O., & Schankerman, M. (2001). Characteristics of Patent Litigation: A Window on Competition. *The RAND Journal of Economics, 32*(1), 129–151.

Long, C. (2002). Patent Signals. *The University of Chicago Law Review, 69*, 625–679.

Leberecht, T., & Meienberg, F. (2014). *Private Claims on Nature. No to Syngenta's Patent on Peppers*. No Patents on Seeds. https://www.no-patents-on-seeds.org/sites/default/files/2019-01/2014_Brochure_No_to_Syngenta_s_patent_on_peppers%20komp.pdf. Accessed 19 Oct 2020.

Phillips McDougall. (2011). *The Cost and Time Iinvolved in the Discovery, Development and Authorisation of a New Plant Biotechnology Derived Trait*. Crop Life International. https://croplife.org/wp-content/uploads/2014/04/Getting-a-Biotech-Crop-to-Market-Phillips-McDougall-Study.pdf. Accessed 19 Oct 2020.

Mirowski, P. (2012). The Modern Commercialization of Science Is a Passel of Ponzi Schemes. *Social Epistemology, 26*(3–4), 285–310.

Noel, M., & Schankerman, M. (2013). Strategic Patenting and Software Innovation: Strategic Patenting and Software Innovation. *The Journal of Industrial Economics, 61*(3), 481–520.

Patstat (2019, Spring). European Patent Office. https://data.epo.org/expert-services/index.html. Accessed 19 Oct 2020.

Parthasarathy, S. (2017). *Patent Politics: Life Forms, Markets, and the Public Interest in the United States and Europe*. Chicago: University of Chicago Press.

Perzanowski, A., & Schultz, J. (2018). *The End of Ownership: Personal Property in the Digital Economy*. Boston: MIT Press.

Ragonnaud, G. (2013). *The EU Seed and Plant reproductive Material Market in Perspective: A Focus on Companies and Market Shares*. Brussels: Policy Department B: Structural and Cohesion Policies. https://www.europarl.europa.eu/RegData/etudes/note/join/2013/513994/IPOL-AGRI_NT(2013)513994_EN.pdf. Accessed 19 Oct 2020.

Sanderson, J. (2017). *Plants, People and Practices: The Nature and History of the UPOV Convention*. New York: Cambridge University Press.

Schankerman, M., & Pakes, A. (1986). Estimates of the Value of Patent Rights in European Countries During the Post-1950 Period. *The Economic Journal*, *96*(384), 1052.

Schenkelaars, P., de Vriend, H., & Kalatzaidonakes, N. (2011). *Drivers of Consolidation in the Seed Industry and Its Consequences for Innovation*. Bilthoven: COGEM.

Schiebinger, L. (2011). Prospecting for Drugs: European Naturalists in the West Indies. In S.G. Harding (Ed.), *The Postcolonial Science and Technology Studies Reader* (pp. 110–126). Durham: Duke University Press.

Schubert, J., Böschen, S., & Gill, B. (2011). Having or Doing Intellectual Property Rights? Transgenic Seed on the Edge Between Refeudalisation and Napsterisation. *European Journal of Sociology*, *52*(1), 1–17.

Slaughter, S., & Rhoades, G. (2004). *Academic Capitalism and the New Economy: Markets, State, and Higher Education*. Baltimore: Johns Hopkins University Press.

Torrisi, S., Gambardella, A., Giuri, P., Harhoff, D., Hoisl, K., & Mariani, M. (2016). Used, Blocking and Sleeping Patents: Empirical Evidence from a Large-Scale Inventor Survey. *Research Policy*, *45*(7), 1374–1385.

Van Overwalle, G. (2017). Patent Pools and Clearinghouses in the Life Sciences: Back to the Future. In D. Matthews & H. Zech (Eds.), *Research Handbook on Intellectual Property and the Life Sciences* (pp. 304–335). Cheltenham: Edward Elgar.

Open Access This chapter is licensed under the terms of the Creative Commons Attribution 4.0 International License (http://creativecommons.org/licenses/by/4.0/), which permits use, sharing, adaptation, distribution and reproduction in any medium or format, as long as you give appropriate credit to the original author(s) and the source, provide a link to the Creative Commons license and indicate if changes were made.

The images or other third party material in this chapter are included in the chapter's Creative Commons license, unless indicated otherwise in a credit line to the material. If material is not included in the chapter's Creative Commons license and your intended use is not permitted by statutory regulation or exceeds the permitted use, you will need to obtain permission directly from the copyright holder.

5

Bioenergy, Thermodynamics and Inequalities

Larry Lohmann

5.1 Introduction

This chapter takes a step back from the empirically detailed studies of the bioenergy boom in Latin America, Asia and Europe that comprise the bulk of this book in order to focus on some of the underlying historical dynamics of bioenergy. It does so out of the conviction that only an activism that takes account of the exploitation and appropriation common to specific instances of bioenergy development is likely to be effective in the long term against the degradations and threats to survival that it entails.

The chapter concentrates mainly on the contradictions of the thermodynamic energy developed in the nineteenth century as a background for sketching how bioenergy perpetuates and accentuates these contradictions, and what the consequences are for biofuel developers, energy

---

L. Lohmann (✉)
The Corner House, Sturminster Newton, UK
e-mail: larrylohmann@gn.apc.org

transition enthusiasts and bioenergy critics. Interleaved with this exposition are reflections on how social movements might place themselves more strategically in bioenergy struggles.

5.2 Thermodynamic Energy as Politics

There is little point in studying bioenergy without some idea of what it is. Clearing away some common confusions, anachronisms and teleologies is crucial at the start.

The biggest confusions are around energy itself, not just bioenergy. These confusions can be found in the writings of many respected contemporary historians such as E. A. Wrigley (2010), Rolf Sieferle (2010), J. R. McNeill (Steffen et al. 2011), Kenneth Pomeranz (2000) and Vaclav Smil (2017). Such writers tend to assume lazily that every society in history has possessed fundamentally the same hunger for greater and greater supplies of an item they call "energy". They seldom define this item or inquire into its history. For example, Wrigley (2010, pp. 42, 44, 191, 205), Sieferle (2010, p. 137) and McNeill (in Steffen et al. 2011, p. 848) each write that an energy "bottleneck" in pre-industrial societies frustrated an intrinsic, pan-human desire for growth—a bottleneck that was only broken with the advent of the fossil fuel era in nineteenth-century Europe.

But there was no such bottleneck (Malm 2016). There was no such energy. The practice and theory of energy, energy stocks and energy sources that we take for granted today did not exist before 1800. At that time, no extensive industrial conversion and transfer infrastructure existed that was capable of uniting and commensurating various thermal, dynamical, electrical and other phenomena into the unitary, indestructible, abstract force that only later came to be called "energy". It was not yet possible even to think about disentangling muscular exertion, wood burning and falls of water (say) from the diverse social or natural contexts in which they were embedded, re-entangle them into systems of exchangeable "equivalents", and accumulate the transformed result into a single pile. Before then, as historian Joel Mokyr (1999, pp. 20–21) observes, "the notion that a horse pulling a treadmill and a coal fire

5 Bioenergy, Thermodynamics and Inequalities

heating a lime kiln were in some sense doing the same thing would have appeared absurd". None of these things were "energy consumption". Energy had neither use-values nor exchange value because there was nothing identifiable as such to be valued. There were no "energy companies". There was no "energy sector", no "energy outlook", no "energy planning", no "energy transition", nor any "energy alternatives". It would have been hard to explain units of measurement like joules, BTUs, kilogramme-metres, ergs, dynes, calories, therms, newtons or barrels of oil-equivalent. Concepts such as "energy return on investment" (EROI) would have been incomprehensible. Although steam engines were already being used early in the eighteenth century for specialized purposes (indeed there had been toy steam engines in ancient Greece), they had not yet begun converting the latent heat of coal stocks into mechanical energy on a scale sufficient to restructure whole industrial, transport and shipping systems. The electric batteries first described in 1800, which transform chemical energy into electrical energy and back, were as yet only a curiosity. Dynamos for converting mechanical energy into electricity did not exist until 1830 or become industrially significant until the 1870s. Electric motors for converting electricity back into mechanical energy appeared only in the 1830s and were embedded into industry only in the 1890s. It was only in the mid-nineteenth century, similarly, that the telegraph began to entrench the mutual convertibility of electricity and magnetism into everyday experience worldwide. Solar cells that could convert sunlight to electricity were not built until 1839. Internal combustion engines for converting chemical to thermal to mechanical energy, although conceived before 1800, began to be marketed only in the 1860s and 1870s.

Correspondingly, it was only between around 1820 and 1850 that today's *concept* of energy (if anything that remains so incoherent can be called a concept) began to take shape via thermodynamics (see Thermodynamics: The First and Second Laws). Thermodynamics developed principally out of the project of a certain privileged group of male Northern Europeans to theorize industrial heat engines (Daggett 2019, p. 37). It was impelled largely by the need to help machines provide business with labour productivity increases, labour discipline, labour concentrations and relative independence from a multitude of ingrained human and more-than-human rhythms, as well as speedier realization of the value of commodities (Malm 2016; Huber 2009). Its "regulative idea" of unified energy also influenced the mapping and organizing of new frontiers for extraction of fuels for capital's conversion devices. To adapt the terminology of Naoki Sakai (1997, p. 41), the growth of energy science, like that of so many other disciplines, was not "determined by the existence of its object". Rather, the object that emerged was "made possible by the existence of the discipline".

> **Thermodynamics: The First and Second Laws**
> In 1865, the great German physicist Rudolf Clausius summarized thermodynamics in two laws:
> 1. The *energy* of the universe is constant.
> 2. The amount of *usable energy* declines (i.e. *entropy* tends to increase in a closed system).
>
> The First Law inspired capitalists to try to put the entire universe to work. It conceptualized a monolithic "energy" that was both inexhaustible and interconvertible. Whatever capital needed to make machines run—mechanical force, heat, electricity, magnetism, light—could be conjured up from any other form of energy that was lying around, given enough ingenuity.
>
> The Second Law revealed the other side of the story. It showed that the more that capital instrumentalized waterfalls, fire, wind, coal, magnetism and so forth as being mere aspects of this great pool of abstract energy, the less of the new energy actually became available for capital's own use. The more that energy was converted back and forth into different forms (Smil 2017, p. 26), the more of it was "degraded". Linear time assumed a new prominence in the shape of an arrow indicating a one-way trip towards universal "heat death".
>
> The contradiction between the two laws reflects the contradictions of the capitalist society that gave rise to them. The First Law helped capital treat the world as a limitless, fungible resource. The Second Law exposed the flip side: waste, pollution and disorder that would ultimately cripple industrial capital itself.

5 Bioenergy, Thermodynamics and Inequalities

Yet even after the First Law began to be formulated, energy was still not treated as a single abstract fluid that could be transferred in large quantities over long distances. The first articulations of the Law set up methodologies for calculating equivalences among previously separated domains like motion and heat, but did not mention conservation or transformation of a singular energy. For example, James Joule, the brewing capitalist who in the 1840s struggled to fix a mechanical "equivalent" for heat, "never said that all forces are essentially differing manifestations of the same ontological 'thing'" (Mirowski 1989, p. 42). That reification was introduced by Lord Kelvin in 1851. Even at the turn of the twentieth century, textbooks were still presenting the First Law more as a "principle of equivalence" than as the "principle of conservation of energy" (Coelho 2009, p. 2651). To be sure, the latter usage is now the popular one (Mirowski 1989, p. 13). Yet the same energy science that has encouraged the public to fetishize energy as an object has also, paradoxically, continually undermined the fetish. By the 1880s, for instance, Hermann von Helmholtz found himself "in cautious retreat from the conception of energy as a mechanistic substance" (ibid., p. 47). In 1918, Noether's Theorem, developed by the Göttingen mathematician Emmy Noether, "drove another nail into the coffin of energy as a substance", cementing a sense that "energy was not really any one thing, but rather a flexible means of expressing symmetry principles" (ibid., p. 72). In 1943, the Harvard physicist P. W. Bridgman argued that energy was just too "hybrid" for it to be possible to set up a "parallelism" between it and "ordinary material things" (Bridgman 1943, p. 115). In the 1960s, the Nobel physicist Richard Feynman famously reminded his audience that the First Law had never actually described "anything concrete": "we have no knowledge of what energy *is*" (Feynman 2010, volume 4, chapter 1). By 2011, it was possible for one prominent thinker to define energy simply as a "relationship of difference that tends to eliminate itself", a "gradient across which there is a tendency to even out and dissipate"— that is, as a *form* rather than a *substance* (Deacon 2011, pp. 218–219). Nothing could be further from the technocratic picture of energy as a universal fuel with sources dotted around the landscape. Yet in a sense, the conception of energy as form rather than substance has been present from the very beginning of thermodynamics. It derives ultimately from the inspired analogy explored by the early nineteenth-century French thermodynamicist Sadi Carnot between the "falls" (*chutes*) of water that riverside mills captured and the "falls" from hot to cold that made heat engines work (Carnot 1988 [1824]). Although Carnot himself thought of heat as a substance, his metaphor paved the way for a conception of energy that emphasized relations and terrains rather than magical stuff.

Throughout its early development, energy science was created largely by engineers who shared the interests of business. "An *economic* point of view formed the root of thermodynamics", historian Theodore Porter (1994, p. 141) emphasizes. "Economic and physical ideas grew up together, sharing a common context". As historians of science Crosbie Smith and M. Norton Wise (1989, pp. xx–xxi) note, the mathematical physics of Lord Kelvin was "thoroughly permeated" by an industrial capitalist cosmovision. Nor would energy as we know it have come into being without empire, for which thermodynamics was a welcome way of reorganizing activities belonging to diverse networks of life into a monolithic, unitary "energy"-supporting worldwide deployment of steam engines, vortex turbines and transoceanic cables. Indeed, the emergence of coal as an imperial "fossil fuel" itself played a part in cementing thermodynamic energy into political thinking worldwide. Simply because it was by nature abundant, concentrated, easily transportable and accumulable, usable in a wide variety of contexts, and completely independent of annual plant growth cycles, coal (and later, oil and gas) was an avatar for the kind of "universal energy equivalent" that thermodynamics was already advocating. It was not that there had always existed a primordial pan-human need for more and more abstract "energy" that fortunately just happened to be relieved one day by fossil fuels. Rather, fossil fuels themselves helped shape the idea of such an energy. Of course, for millions of years, long before *Homo sapiens* came along, processes had been going on that were later described as the transformation of sunlight into thermal energy, thermal and chemical into mechanical energy, mechanical energy into electricity and so forth. None of that began in the nineteenth century. But without post-1800 imperial reorganizations of relations among humans and nonhumans, energy and its ideology could never have become so hegemonic.

If capital and empire made thermodynamics what it is, thermodynamics fully returned the favour (Daggett 2019). Thermodynamic energy was there to carry out an unlimited amount of what early nineteenth-century Cornwall engineers measured as the "duty" done by given quantities of coal (Cardwell 1993, p. 117). It was pictured as being available to business anywhere on earth or off it. "Energy sources" became geographic features, whether they were rivers with steep drops, coal seams, oil fields, peninsulas with steady high winds, uranium deposits, deserts with high rates of insolation, forests with chippable trees or stretches of soil suitable for oil palm, jatropha or sugar cane. Mapped in this way, energy sources could be made to overlie, overlap or even obliterate other geographical features such as cultivated land, indigenous territories, water sources, grazing grounds or customary property or political boundaries. The abstract energy symbolized in the First Law of Thermodynamics became a real part of the world, entangling itself into the emergent "abstractified" forms of work, society, space, water and nature that to some extent had preceded it on the early modern landscape. Thermodynamics was thus a crucial technique of enclosure of commons across Europe, India, Africa and the Americas. It was also indispensable to the mobilization of millions of newly landless labourers and slaves in centres of mechanization, as well as to the growth of the role of commodity exchange in providing the necessities of life.

Today's thermodynamically rationalized "biofuel complex", for instance, has engendered a new concept of "marginal land" in the tradition of older colonial notions of "waste" and *terra nullius*. Areas of land identified through remote sensing as "non-competitive" for purposes of industrial food production become acceptable sites for energy extraction on a par with deposits of peat or oil shale. Like the twentieth-century notion of "sacrifice zones" (National Academy of Sciences 1974), this geographic/thermodynamic methodology tends to obscure many other features of the land in question (Nalepa and Bauer 2012). These include not only the capacity to supply medicine, provide building materials and sustain hunting, gathering, grazing and subsistence farming. They also include the plural non-thermodynamic or "little-e" energies (Lohmann and Hildyard 2014) inherent to the territory, maintaining the mutual incommensurability of which is often central to local livelihood strategies and resistance to encroachments of big capital and the state.

In addition to augmenting a notion of efficiency that has contributed to modern racism and colonialism (Daggett 2019; Alexander 2008), the energy abstraction has also significantly expanded the domain of global scarcity. Until firewood from a common woodland becomes "energy", it is not necessarily scarce in an economic sense (Lohmann and Hildyard 2014, pp. 63–64). Correspondingly, it is only with thermodynamic energy that modern waste really comes into its own, as well as the types of human labour that are needed to help clean up, stow, reuse, manage, absorb or hide it. Thermodynamic energy is all about putting together diverse Carnotian "falls" or irreversible erosions of difference across broad geographies. As such, it unavoidably generates forms and volumes of "waste" energy specific to the age of the Second as well as the First Laws of Thermodynamics. This is so whether the "falls" in question consist of water sluicing through dam penstocks, air impacting on the blades of windmills, "falls" of electrons through the electrolyte of batteries connected to closed circuits, "falls" from hot to cold within aircraft engines, or "falls" of soil fertility into sterility on industrial biomass plantations. Water can't be returned to the height from which it falls using only its own energy, nor ash, heat and CO_2 reassembled into coal. That makes it all the more imperative to find cheap ways of clearing detritus out of the way so that costs can be saved. Any locomotive that is to go on pulling railway coaches has to be provided with a place to vent its smoke, and workers to scrub it free of soot. Every Google translation machine needs dedicated, cheap living human or non-human activity to clear away the debris it generates, whether it be carbon dioxide or inappropriate word sequences.

But usable energy can become "waste" also just by being allowed to lie around unused. For example, the "falls" of river systems that are not exploited with hydroelectric dams that "break" the falls become viewed by technocratic organizations such as the Mekong River Commission as "wasted resources", much to the bemusement of communities that have depended on them for generations. Other wastes emerge after the dams are built and the reservoirs behind them silt up, removing the "fall" and necessitating dredging that must be powered via the exploitation of further "falls" elsewhere. Even the space that a quiet lake occupies can suddenly start looking like "waste" when the lake's energy is revealed to be "unusable". Insofar as thermodynamic energy amounts to a running modification of landscapes to exploit "falls" in pursuit of a good structured as unlimited (Hornborg 2001), it is also a shorthand for the unlimited expansion of the frontiers of degradation. It is not so much that the need for thermodynamic energy necessitates the physical and political re-engineering of territories. In a more elegant formulation, thermodynamic energy *is* the re-engineering of territories.

This "denaturalization" of the history of energy (Bonneuil and Fressoz 2016, p. 64) helps us understand thermodynamic propositions not only as valid science, but also as chunks of political ideology. For example, as the literary historian and physicist Barri Gold (2010, p. 9) observes, the Second Law idea that the amount of energy "beyond our use" tends to increase in a closed system can't escape questions such as: "What use? And who's *we*?" That generically anthropomorphic "we" (Georgescu-Roegen 1975, p. 351) would be unlikely to include indigenous peoples for whom what physicists would call the "unavailable" or "disordered" energy in a calm lake is far from useless, or peasants who deny that the energy in a fast-flowing river is "wasted" unless converted into hydro-electricity. The reality, indeed, is that thermodynamics tends to hide a vast, churning, enduring "underground" of anti-thermodynamic energies. These energies are around us always and everywhere, in cities as well as rural areas, in hospitals and factories as well as irrigation systems. Examples include the growth of vegetables in contemporary urban gardens in Milwaukee; the burning of commons firewood in Chiang Rai; or the bubbling of springs in the water-hill-village systems of Totonac communities in the Sierra Norte de Puebla (Smith 2007). The logic of such energies militates against aggregating them with one another and disentangling them from the limited goods of subsistence in the manner of the First Law of Thermodynamics. It also militates against becoming overly preoccupied with the Second Law, whose barrier to a notional "ideal efficiency" is more of an object of dread to capitalist technocrats and ecological modernizers than to ordinary people.

For millions of individuals virtually all of the time, and for everybody at least some of the time, periodically refusing thermodynamic energy's claim to be able to subsume anti-thermodynamic energies into itself is just a matter of common sense and survival. So, too, is keeping at arm's length thermodynamic projects featuring relentlessly mounting levels of disorder and waste that require professional management and concealment. Many rural areas worldwide reveal living retorts to such projects in the form of homely ways of working the vernacular wastes of commons—as when food waste is integrated into animal-raising, animal waste into field care and plant waste, cleaned up and recycled through fire, into the care of grain, forests, water and humans alike. From a subsistence perspective, such practices are not "renewable energy". They are not energy at all. They do not exemplify efficiency and are not productive of anything except themselves.

In fact, it is only when the heat source (the frontier of "usable energy") for the industrial engines that thermodynamics has worked to improve becomes a real abstraction that the heat sink (the zone of degraded energy or high "entropy") becomes a capitalist obsession and a global environmental issue. An ordinary rural community striving to take care of a local stream that never runs dry is typically not preoccupied with Wilhelm Ostwald's thermodynamically inspired "energetic imperative"—"do not waste any energy, make it useful" (Ostwald 1912, p. 85)—nor Nicholas Georgescu-Roegen's (1971) similarly motivated Malthusian cautions. For such a community, the Second Law is not necessarily a problem, any more than a perpetual motion machine that overcomes the Second Law is a shimmering "sublime object of ideology" (Žižek 1989). Nor would there be any point in praising such a community for "efficiency". Efficiency as understood today, riven by some of the same deep contradictions that afflict capital and thermodynamics (Polimeni et al. 2008), derives from a different context, that of industrial machines and their interpolation into societies organized around limitless accumulation.

The irony is—and this is an insight that is unfortunately missing from nearly all current global energy and climate debates, including debates over bioenergy and the so-called energy transition—that anti-thermodynamic energies are not only ubiquitous, but also, paradoxically, essential to maintaining the precarious status of thermodynamic energy itself. Together, the two form a "contradictory unity" (Harvey 2014) analogous to those that uneasily link living with dead labour, use value with exchange value, unpaid reproductive work with wage labour, and commons with capitalist forms of socio-natural organization. Unavoidably, "inside" the monolith of official energy can always be found a "hidden abode" (Fraser 2014) inhabited by a plurality of vernacular energies that it converts, commensurates, parasitizes, degrades and exhausts, yet which through their very opposition make it possible to accumulate surplus value (Toscano 2018). To capitalist planners, for example, a common woodland may at first sight look like either raw material for thermodynamic energy or an obstacle to be eliminated to make way for a hydroelectric dam, oil refinery or wind farm that produces more of it. Yet when economic crises hit and the planners themselves face redundancy, they may suddenly find themselves "recognizing" the woodland's non-thermodynamic energies as a useful zero-cost subsidy that helps maintain local workers pending their re-employment in the service of machines powered by thermodynamic energy. Or the planners may find themselves paradoxically dependent on the creative subversion that commoners exercise by thieving grid electricity to sustain subsistence systems dedicated to thwarting the commensuration of "little-e" energies into thermodynamic energy. In practice, it is only in conjunction with non-energetic, non-entropic or "negentropic" (Schrodinger 1944) enclaves in commons and elsewhere that the massively energetic, massively entropic machines reliant on thermodynamic energy become capable of working for capital for any significant length of time.

To put it another way, thermodynamic energy did not emerge once and for all in the nineteenth century. It *continues* to emerge in tandem with frontiers of resistance to its dominance. With a bit of patience, this resistance can be recognized in every kitchen, back garden, slum and factory floor. It can be glimpsed in every social movement fighting mining operations or even just local rights of way that highways catering to internal combustion engines threaten to break up. Thermodynamic energy is always under construction, but also always being undone, in millions of locations. Struggles contesting it form one part of continuing battles against enclosure of all kinds, including the dominance of the concept of resources. Any political struggle whose horizons extend beyond correcting prices, improving efficiency and securing wage work towards confronting patriarchy, racism, coercive capitalist social relations, commodity fetishism and capitalist work itself is likely eventually to find itself joining in already-existing movements confronting the hegemony of thermodynamic energy (Ediciones Inéditos 2019; Daggett 2019). So will any climate movement that seeks to build solidarity with workers, peasants and indigenous peoples struggling against oil extraction or bioenergy plantations instead of trying somehow to ally itself with "physics" against a generic class of human carbon dioxide emitters (Davis 2019; Invernizzi-Accetti 2019). As Christophe Bonneuil and Jean-Baptiste Fressoz (2016, p. 63) emphasize, any serious response to the "shock of the anthropocene" will need, in a sense, to "free itself from […] the very concept of energy" and the "project that brings every form of work (from brain to blast furnace) into a generalized equivalence".

Like many other political projects of "masterful" abstraction, thermodynamic energy has a particular gender (Lutz 1995), a particular racial "colour" (Eze 1997; Dabashi 2015) and a particular class. It bears a bias against the practices of many oppressed groups accustomed to showing respect for a fire, a stream or a tool as "one of us" (Lenkersdorf 2008). And it sets its face against societies for whom "our history is the future" (Estes 2019) insofar as it superimposes the one-dimensional arrow of time of the Second Law of Thermodynamics on spiral or multi-dimensional time (Cusicanqui 2015), in which present and past events can be simultaneous or mutually embodied in one another (Anderson 2006, pp. 22–36). Like the frequent white feminist failure to interrogate race, or the common failure of antiracism to interrogate patriarchy (Crenshaw 1989), any failure of liberation movements to interrogate thermodynamic oppression is bound to reinforce the subordination and unequal status of peasants, workers, women, indigenous peoples and the colonized everywhere.

5.3 Bioenergy as Thermodynamic Energy: Deepening the Contradictions

As is obvious from the other chapters in this book, adding the prefix "bio-" to energy changes nothing about its essential ecological and political characteristics. Bioenergy—in the sense used in this volume—is not an uncommensurating of the "little-e energies" referred to above nor a re-embedding of them in diverse commons practices. Nor is it a revalorization of non-energy-mediated relations among human beings and the more-than-human. That path is blocked by the angel with the flaming sword. Instead, bioenergy is thermodynamic energy and remains subject to all its contradictions. Opening a new chapter in the co-evolution of fossil-fuel dependence and thermodynamics, bioenergy is provoking a new phase of the same anti-colonial and anti-capitalist struggles that were modified so decisively by the development of energy itself.

Bioenergy's challenge to coal, oil and gas, in short, is purely notional. Bioenergy demands that living biomass supplement and substitute for fossil biomass as precisely the same kind of "universal fuel" that thermodynamics helped make possible. Its claim to be a fossil-free thermodynamic energy is a delusional denial of that energy's very fossil inheritance. Far from confining itself to enlisting living plant matter to round out the low-cost self-provisioning of reserve and other armies of labour, bioenergy policy indeed jams it ever more forcibly down onto the painful Procrustean bed of industrial capital's thermodynamic abstractions. Four hundred times more forcibly, in fact, given that capital has long been committed to appropriating the thermodynamic "equivalent" of at least 400 years of current plant growth in the form of fossil fuels for every year it continues to exploit human labour (Dukes 2003), and is now asking living biomass to help it, *per impossibile*, to strive for the same objective.

Take, for instance, the aviation industry, which is attempting to treat living biomass as if it were fossil fuel in two different ways. First, aviation biofuels are supposed to be able to "substitute" for kerosene, in spite of the fact that an area of land equivalent to that of a medium-sized country would have to be found and permanently set aside to grow plant fuels thermodynamically capable of replacing aviation's share of world petroleum consumption, and in spite of the tremendously entropic follow-on effects. Second, fossil carbon emissions from aviation are supposed to be able to be "offset" by yearly plant growth under the Carbon Offsetting and Reduction Scheme for International Aviation (CORSIA). This would require the annexation of land areas of the same order of magnitude again, together with the maximal formal and real subsumption of what is now called the "organism performance" of the most carbon-productive plant species and the ecosystems and human communities that are recrafted, degraded and progressively exhausted for the sake of their cultivation. This intensified dynamic of thermodynamic enclosure is likely to have similar outcomes at the grassroots whether it is impelled by environmental regulation or by efforts to transform biofuels, carbon offsets or biopatents into commodities, assets, claims on rent or objects for financial speculation—or by all of the above.

Arguably, that places today's battles against bioenergy projects at the very forefront of the two-century-old struggle resisting the dominion of thermodynamic energy. The experience of—for example—the Indonesian oil palm plantation worker in Malaysia or Indonesia (see Janina Puder and Hariati Sinaga in this volume) lies at the intersection (Crenshaw 1989) of multiple oppressions: landlessness, subjection to machine discipline, precaritization, externalization of reproduction costs, nation, patriarchy, race, class—but now also the expanded hegemony of nineteenth-century thermodynamic energy, as increasing amounts of biomass are pressed into service as substitutes for energy-dense hydrocarbons. The Indonesian woman migrant in Malaysia cannot be spoken for by the male plantation worker, nor by the Indonesian peasant woman, nor by the formally educated Northern critic of thermodynamics, nor by a committee of the three. Yet while only she can say "when and where she enters" onto a path of liberation (ibid., p. 160), it is also true that only with her entry along that path that *others* struggling with thermodynamic energy can also enter.

In their analysis of the "biofuel delusion", Mario Giampietro and Kozo Mayumi (2009) conclude that

> we do not need alternative energy sources to keep alive an obsolete pattern of economic growth. What we need is an alternative pattern of

development that will make it possible to use alternative energy sources. (pp. 256–257)

This chapter has sought to go a step or two further still, by exposing the contradictions inherent in the very act of treating cane ethanol, wood pellets or aviation biofuel as energy, as well as by asking whether it is worth even talking about an "energy transition" that does not challenge the dominance of thermodynamics itself. A strategically effective critique of bioenergy developments has to go all the way down into energy itself.

One last way of summarizing the lesson of this chapter is to note the sense in which all bioenergy developments *inherently* entail the intensification of what this book's title refers to as "global inequalities". At a time when debates about energy equality are still overwhelmingly concerned only about the *distribution* of energy and of the costs of its production and circulation, it is more important than ever to stress that the prior issue is actually the *constitution* of energy. In a sense, inequalities are what thermodynamic energy is *for*. It should surprise no one that access to electricity, heat and motive power remains so skewed throughout the world, nor that so many have to suffer to make energy available to so few. From its beginning, thermodynamics has been a mode of denying practices that hundreds of millions of people depend upon to flourish and subsist. It is an integral part of a political settlement achieved and precariously maintained since the nineteenth century under which frontiers of appropriation are organized to make it possible for fossil capital to go on getting something for nothing from a commons both human and extra-human (Moore 2015). Global inequalities connected with bioenergy development and energy transitions can't begin to be seriously addressed without addressing energy itself.

Acknowledgements Thanks for help to Simon Pirani, Nick Hildyard, Ivonne Yanez, Maria Backhouse, Fabricio Rodríguez and the Junior Research Group on Bioeconomy and Inequalities at Friedrich Schiller University, Jena, where another version of this paper was presented.

References

Alexander, J.K. (2008). *The Mantra of Efficiency: From Waterwheel to Social Control.* Baltimore: Johns Hopkins University Press.

Anderson, B. (2006). *Imagined Communities* (revised edition). London: Verso.

Bonneuil, C., & Fressoz, J.-B. (2016). *The Shock of the Anthropocene: The Earth, History and Us.* (Trans.: Fernbach, D.). London: Verso.

Bridgman, P.W. (1943). *The Nature of Thermodynamics.* Cambridge: Harvard University Press.

Cardwell, D.S.L. (1993). Steam Engine Theory in the 19th Century: From Duty to Thermal Efficiency; from Parkes to Sankey. *Transactions of the Newcomen Society, 65*(1), 117–128.

Carnot, S. (1988 [1824]). *Reflections on the Motive Power of Fire* (Trans.: Thurston, R.H.). Ed. with an Introduction by Mendoza, E. Mineola: Dover.

Coelho, R.L. (2009). On the Concept of Energy: History and Philosophy for Science Teaching. *Procedia Social and Behavioral Sciences, 1,* 2648–2652.

Crenshaw, K. (1989). Demarginalizing the Intersection of Race and Sex: A Black Feminist Critique of Antidiscrimination Doctrine, Feminist Theory and Antiracist Politics. *University of Chicago Legal Forum, 1,* 138–167.

Cusicanqui, S.R. (2015). *Sociología de la imagen: Miradas ch'ixi desde la historia andina.* Buenos Aires: Tinta Limón.

Dabashi, H. (2015). *Can Non-Europeans Think?* London: Zed Books.

Daggett, C.N. (2019). *The Birth of Energy: Fossil Fuels, Thermodynamics and the Politics of Work.* Durham: Duke University Press.

Davis, D. (2019, April 16). Climate Change Is 'Greatest Challenge Humans Have Ever Faced,' Author Says. *National Public Radio.* https://www.npr.org/2019/04/16/713829853/climate-change-is-greatest-challenge-humans-have-ever-faced-author-says. Accessed 12 Jan 2020.

Deacon, T.W. (2011). *Incomplete Nature: How Mind Emerged from Matter.* New York: Norton.

Dukes, J.S. (2003). Burning Buried Sunshine: Human Consumption of Ancient Solar Energy. *Climatic Change, 61,* 31–44.

Ediciones Inéditos (2019). Prole Wave: Climate Change, Circulation Struggles and the Communist Horizon. *Non.copyriot.com.* https://non.copyriot.com/prole-wave-climate-change-circulation-struggles-the-communist-horizon/. Accessed 12 Jan 2020.

Estes, N. (2019) *Our History Is the Future: Standing Rock Versus the Dakota Access Pipeline, and the Long Tradition of Indigenous Resistance*. London: Verso.

Eze, E.C. (1997). The Color of Reason: The Idea of "Race" in Kant's Anthropology. In E.C. Eze (Ed.), *Postcolonial African Philosophy: A Critical Reader* (pp. 103–140). Oxford: Blackwell.

Feynman, R. (2010). *The Feynman Lectures on Physics*. Vol. 1. New York: Basic.

Fraser, N. (2014). Behind Marx's Hidden Abode: For an Expanded Concept of Capitalism. *New Left Review*, 86, 55–72.

Georgescu-Roegen, N. (1975). Energy and Economic Myths. *Southern Economic Journal*, 41(3), 347–381.

Georgescu-Roegen, N. (1971) *The Entropy Law and the Economic Process*. Cambridge: Harvard University Press.

Giampietro, M., & Mayumi, K. (2009). *The Biofuel Delusion: The Fallacy of Large-Scale Agro-Biofuel Production*. London: Earthscan.

Gold, B.J. (2010). *ThermoPoetics: Energy in Victorian Literature and Science*. Cambridge: The MIT Press.

Harvey, D. (2014). *Seventeen Contradictions and the End of Capitalism*. Oxford: Oxford University Press.

Hornborg, A. (2001). *The Power of the Machine: Global Inequalities of Economy, Technology and Environment*. Walnut Creek: Altamira.

Huber, M. (2009). Energizing Historical Materialism: Fossil Fuels, Space and the Capitalist Mode of Production. *Geoforum*, 40(1), 105–115.

Invernizzi-Accetti, C. (2019, December 29). Climate Change Denial May Have Been Defeated in 2019, but What Comes Next Won't Be Easier. *The Guardian*. https://www.theguardian.com/commentisfree/2019/dec/29/the-climate-movement-is-about-to-get-more-political-and-thats-a-good-thing. Accessed 22 Oct 2020.

Lenkersdorf, C. (2008). *Aprender a escuchar: Enseñanzas maya-tojolabales*. Mexico D.F.: Plaza y Valdés.

Lohmann, L., & Hildyard, N. (2014). *Energy, Work and Finance*. Sturminster Newton: The Corner House. http://www.thecornerhouse.org.uk/sites/thecornerhouse.org.uk/files/EnergyWorkFinance%20%282.57MB%29.pdf. Accessed 22 Oct 2020.

Lutz, C. (1995). The Gender of Theory. In R. Behar & D.A. Gordon (Eds.), *Women Writing Culture* (pp. 249–266). Berkeley: University of California Press.

Malm, A. (2016). *Fossil Capital: The Rise of Steam Power and the Roots of Global Warming*. London: Verso.

Mirowski, P. (1989). *More Heat Than Light: Economics as Social Physics, Physics as Nature's Economics.* Cambridge: Cambridge University Press.

Mokyr, J. (1999). Editor's Introduction: The New Economic History and the Industrial Revolution. In J. Mokyr (Ed.), *The British Industrial Revolution: An Economic Perspective* (pp. 1–127). Boulder: Westview.

Moore, J. (2015). *Capitalism in the Web of Life: Ecology and the Accumulation of Capital.* London: Verso.

Nalepa, R.A., & Bauer, D.M. (2012). Marginal Lands: The Role of Remote Sensing in Constructing Landscapes for Agrofuel Development. *Journal of Peasant Studies, 39*(2), 403–422.

National Academy of Sciences [US] (1974). *Rehabilitation Potential of Western Coal Lands: A Report to the Energy Policy Project of the Ford Foundation.* Cambridge: Ballinger.

Ostwald, W. (1912). *Der energetische Imperativ.* Leipzig: Akademische Verlagsgesellschaft.

Polimeni, J.M., Mayumi, K., Giampietro, M., & Alcott, B. (2008). *The Jevons Paradox and the Myth of Resource Efficiency Improvements.* London: Earthscan.

Pomeranz, K. (2000). *The Great Divergence: China, Europe, and the Making of the Modern World Economy.* Princeton: Princeton University Press.

Porter, T.M. (1994). Rigour and Practicality: Rival Ideals of Quantification in Nineteenth-Century Economics. In P. Mirowski (Ed.), *Natural Images in Economic Thought* (pp. 128–170). Cambridge: Cambridge University Press.

Sakai, N. (1997). *Translation and Subjectivity: On Japan and Cultural Nationalism.* Minneapolis: University of Minnesota Press.

Schrodinger, E. (1944). *What Is Life?* Cambridge: Cambridge University Press.

Sieferle, R. (2010 [1982]). *The Subterranean Forest: Energy Systems and the Industrial Revolution.* Cambridge: The White Horse Press.

Smil, V. (2017). *Energy and Civilization: A History.* Cambridge: The MIT Press.

Smith, C., & Wise, M.N. (1989). *Energy and Empire: A Biographical Study of Lord Kelvin.* New York: Cambridge University Press.

Smith, W.D. (2007). Presence of Mind as Working Climate Change Knowledge: A Totonac Cosmopolitics. In M. Pettenger (Ed.), *The Social Construction of Climate Change: Power, Knowledge, Norms, Discourses* (pp. 217–234). Aldershot: Ashgate.

Steffen, W., Grinevald, J., Crutzen, P., & McNeill, J. (2011). The Anthropocene: Conceptual and Historical Perspectives. *Philosophical Transactions of the Royal Society A, 369,* 842–867.

Toscano, A. (2018). Antiphysis/Antipraxis: Universal Exhaustion and the Tragedy of Materiality. In B.R. Bellamy & J. Diamanti (Eds.), *Materialism and the Critique of Energy* (pp. 471–499). Chicago: MCM Publishing.

Wrigley, E.A. (2010). *Energy and the English Industrial Revolution*. Cambridge: Cambridge University Press.

Žižek, S. (1989). *The Sublime Object of Ideology*. London: Verso.

Open Access This chapter is licensed under the terms of the Creative Commons Attribution 4.0 International License (http://creativecommons.org/licenses/by/4.0/), which permits use, sharing, adaptation, distribution and reproduction in any medium or format, as long as you give appropriate credit to the original author(s) and the source, provide a link to the Creative Commons license and indicate if changes were made.

The images or other third party material in this chapter are included in the chapter's Creative Commons license, unless indicated otherwise in a credit line to the material. If material is not included in the chapter's Creative Commons license and your intended use is not permitted by statutory regulation or exceeds the permitted use, you will need to obtain permission directly from the copyright holder.

Part III

Bioeconomy Policies and Agendas in Different Countries

6

Knowledge, Research, and Germany's Bioeconomy: Inclusion and Exclusion in Bioenergy Funding Policies

Rosa Lehmann

6.1 Introduction: Bioenergy's Uncertain Prospects

The future of Germany's bioenergy is unclear. Bioenergy is commonly associated with rapeseed and corn monocultures, and with wood chip heating or biogas plants, which turn either scrap wood or cow manure and cultivated biomass into electricity and heating, respectively. The research into bioenergy production by private farmers or (and this is particularly the case in Germany) bioenergy cooperatives has a firm place in the growing body of social science literature on the global energy transition. These citizen-based renewable energy projects serve as examples of a decentralized energy transition, i.e. a transition where energy is produced and consumed not centrally but in the rural areas or the neighbourhoods in which it is produced. This anchors the energy transition socially and contributes to its success (see Kunze 2012; Morris

R. Lehmann (✉)
Institute of Sociology, Friedrich Schiller University Jena, Jena, Germany
e-mail: rosa.lehmann@uni-jena.de

and Jungjohann 2016; Radtke 2016). In studies that explicitly adopt an energy justice perspective (Jenkins et al. 2016), these models increase participation and benefit-sharing due to ownership structures that enable a broader scale of involvement than is possible with large private stock companies (e.g. Kunze 2012; Szulecki 2018).[1] Although different studies and experts predict the survival of bioenergy production (see Szarka et al. 2017; Strzalka et al. 2017), recent research reveals that the social dimension of bioenergy production as described above is facing uncertainties concerning regulatory and technological developments, such as the expiration of subsidies and the need for the flexibilization of biogas plants (Backhouse et al. 2020).

Discussions about the technological aspect of bioenergy in the energy transition centre on bioenergy technology and innovative biomass. The future of the energy system is said to rely on "the artificial leaf" in terms of artificial photosynthesis (Marshall 2014), on algae and other microorganisms in energy generation, e.g. on building walls, or fuels produced from straw residues and cup plants, with the latter turning erosion-prone slopes into flourishing, bee-friendly landscapes.[2] One starts to imagine algae tanks on the roofs of public libraries and futuristic artificial bushes replacing green-white biogas plants. Nevertheless, the social dimension of these technological visions is somewhat blurred.

The German Bioeconomy Strategy began with the publication of a document in 2010 by the Federal Ministry of Education and Research (BMBF 2010). The Bioeconomy Strategy is focused on future technologies and research and innovation (R&I) and constitutes a puzzle for the attempt to assess the prospects of bioenergy and, most notably, its social dimension. Research in science and technology studies (STS) and political ecology has shown that the use of a resource or technology (for energy production) reshapes social relations (see Görg 2004;

---

[1] It is noteworthy that "civil society" is neither a homogenous actor, nor are these models 100% inclusive (see Radtke 2016).

[2] See, e.g., the recommendation of the German Bioeconomy Council (German Bioeconomy Council 2016) or the contributions on the website https://www.pflanzenforschung.de, which represent findings of research funded by the Federal Ministry of Education and Research (BMBF) on applied plant research.

Huber 2015; Lohmann and Hildyard 2014; Miller et al. 2013). Consequently, I understand energy systems as socio-technical systems that not only comprise technology and raw materials, but also human labour, economic investments, institutions, norms, narratives, and power asymmetries between unequally included social groups (Miller et al., p. 136). Thus, I refer to the concept of socio-energy systems in this contribution.

Bioenergy played a role in the Bioeconomy Strategy until 2019. The strategy supports research into the material, chemical, and energetic use of biomass and biological knowledge in order to fundamentally transform the economy. It is embedded in the High-Tech Strategy and in different sustainability and energy policies (see Meyer 2017) and further claims to encompass not only the technological but also the social aspects of the transformation towards a bio-based economy (Fraunhofer ISI 2017, p. 7; BMBF and BMEL 2014; BMBF 2010, 2014; BMEL[3] 2014). However, bearing in mind the importance of decentralized citizen-based bioenergy production for dynamics of the German energy transition, how can we interpret bioenergy-related R&I in the bioeconomy and the associated predictions relating to bioenergy technology? To answer this question, I focus on the Bioeconomy Strategy to deduce the role of bioenergy in R&I and ask how and to what extent bioenergy is related to the production of knowledge in the German strategy. Whose and what kind of knowledge about energy production is supported in funding policies? However, it is also important to understand who and what are excluded when it comes to the social dimension of the current technology-driven bioeconomy (see Birch et al. 2010) as well as for debates about justice in future socio-energy systems.

For the analysis of inclusions and exclusions, I am guided by the proposal of Miller et al. (2013, pp. 136–137) to structure research on energy transitions along the lines of the following analytical categories: energy infrastructures, energy epistemics, and energy justice. I focus on inequalities related to R&I, hence on the latter two categories. I argue that unequal energy epistemics are reflected in the gap between technology-laden research and existing social practices of bioenergy production. The focus of bioeconomy-related R&I is on high-technology

---

[3] Federal Ministry of Food and Agriculture.

innovation and tends to neglect existing experiences and practices of different actors in the bioenergy sector. However, it would be crucial to take the history of socio-(bio)energy systems into account for further biomass-centred strategies that encompass, at least rhetorically, the social dimension of a transition to a bioeconomy.

The research perspective taken here may seem surprising. Technical, governance, and sustainability studies on bioenergy exist for the European context (e.g. Bentsen et al. 2019; Lewandowski 2015; Szarka et al. 2017). Social science research on bioeconomy examines bioenergy in terms of biomass production, sustainability issues, and policies (e.g. Toivanen in this volume) or assesses the perspective of political-institutional changes or/and narratives in the bioeconomy (e.g. Goven and Pavone 2014; Giurca 2018). The shape of R&I in the German Bioeconomy Strategy has been discussed in general assessments (Priefer et al. 2017). STS studies emphasize that an analysis of R&I funding is key to understanding inclusions and exclusions regarding knowledge production (e.g. Frickel et al. 2010; Tyfield et al. 2017). Energy justice literature stresses this dimension for the analysis of renewable energies, although knowledge production is but one category in related research. This chapter contributes to these research fields by examining the shape of bioenergy-related R&I within bioeconomy policies in Germany. It is based on the analysis of grey literature: position papers and press releases, protocols of stakeholder meetings, web pages, press articles, policy and strategy papers, and evaluation reports published by federal ministries and research institutions. Findings are complemented with insights from ten qualitative expert-interviews. The structure is as follows: Sect. 6.2 sketches the analytical framework, Sect. 6.3 presents the preliminary results, and Sect. 6.4 discusses further research.

6.2 Approaching Bioenergy: Epistemics and Justice

As studies on historical and the current energy transition(s) reveal (e.g. Mitchell 2009; Elmhirst et al. 2017), energy systems are inextricably linked to, form, and confine the (unequal) structure of society; hence, they are restructured by changes in the energy system (Miller et al. 2013, p. 136; also Lohmann and Hildyard 2014). Therefore, I refer to them as socio-energy systems. To different extents, research on the social dimension of energy considers the role of knowledge production and the resources actors dispose of to generate expertise and innovation and to promote its realization (historically: Malm 2016). In their introductory article to a journal issue on energy transitions, Miller et al. (2013, pp. 136–137) explicitly apply the role of knowledge production to the exploration of corresponding power relations and inequalities and suggest analysis of energy infrastructures, energy epistemics, and energy justice. In order to examine inclusions and exclusions in bioenergy-related funding policies, the latter two categories are of greater importance for this contribution. Energy epistemics, simply put, is about "[w]ho knows about energy systems, what and how do they know, and whose knowledge counts in governing and reshaping energy futures?" (ibid., p. 137). Studies on energy epistemics and related policies (e.g. Hess and McKane 2017) resonate with findings on knowledge production and inequalities. Along this line, Frickel et al. (2010, p. 467) stress that power relations between actors with different resources shape research priorities and exclude antagonist ideas to a large extent:

> scientific research is increasingly complex, technology-laden, and expensive, [therefore] there is a systematic tendency for knowledge production to rest on the cultural assumptions and material interests of privileged groups. (ibid., p. 446)

Research into the issues that social movements, affected residents, and non-governmental organizations consider important has been, to a lesser degree, funded and completed (ibid.). These include health issues as a consequence of industrial emissions or the use of pesticides in agroindustry (see Arancibia and Motta 2018; Toledo López in this volume). Following Tyfield et al. (2017, pp. 1–9), this is due to the dominant social notion of knowledge as a growing array of "factual, normatively neutral truths" (ibid., p. 2) that result in economic growth. Tyfield et al. claim that the concomitant fetishism of high-technology and research-intensive innovation also applies to research into climate change mitigation and into sustainable substitutes to fossil raw materials (ibid., p. 10). Hence, actors that question this assumption or at least try to balance social and environmental issues with economic concerns are in a less powerful position from the outset. Social scientific research into the bioeconomy confirms that narratives of economic growth dominate over those of nature protection and ecological limits (see Birch et al. 2010; Kleinschmit et al. 2017; Vivien et al. 2019) and that bioeconomy strategies reveal an imbalance through the inclusion of different social groups during agenda setting (e.g. Lühmann 2020; Tittor in this volume).

Energy justice serves as a concept to assess the inclusions and exclusions surrounding energy infrastructures and epistemics. Although grid connections and clean cooking fuels are important when tackling energy poverty and its social and health impacts, as are financial contributions for renewables in attempts to mitigate climate change, energy justice goes beyond these issues to "[s]imply decarbonizing the status quo" (Healy and Barry 2017, p. 457). Due to their materiality, renewables have the potential to be produced, distributed, and consumed in a decentralized energy system (Malm 2016, pp. 37–42). At the same time, they can redistribute ownership and decision-making to different actors, which could prevent the concentration of capital and enable the development of democratized socio-energy systems (Wissen 2016, p. 57). Summing up the energy justice debate, Kirsten Jenkins et al. (2016) put the distributional, recognition-based, and procedural aspects of energy production, distribution, and consumption centre stage: energy justice is aimed at achieving the comprehensive inclusion of actors affected along the value chain (also Avila-Calero 2018; Becker and Naumann 2017; Szulecki 2018; Weis et al. 2015). Here, the equal distribution of costs and benefits is as important as respecting and recognizing the concerns that locals may have about energy projects. Procedural justice refers to transparency and information, equal representation, and the participation of different groups in decision-making. Further, it comprises the role of knowledge of different actors, notably of affected citizens. In the literature, this is often related to Indigenous knowledge. In this chapter, I apply these thoughts to the prospects for bioenergy and current bioenergy-related research in the bioeconomy and argue that (e.g. practitioners', civil society activists') knowledge about decentralized bioenergy production is included in or on the agenda of innovation-driven, research-oriented R&I to much lesser extent.

In sum: inequalities in R&I are of importance for the analysis of transitioning socio-energy systems. A perspective that builds on energy epistemics and justice enables not only the examination of inclusions and exclusions in bioenergy-related knowledge production in the bioeconomy but the possibility of mooting ideas about the social dimension of the energy transition. In the following section, I assess the shape of bioenergy-related R&I in the German bioeconomy.

6.3 Bioenergy in the Transitioning Landscape of the German Bioeconomy: Empirical Insights

Although the focus of this contribution lies on inclusions and exclusions in energy epistemics, existing bioenergy-infrastructures and their role in the current energy transition are of importance for debating and evaluating the character of bioenergy-related R&I in the bioeconomy. They are particularly important for describing the status quo of the production, distribution, and consumption of bioenergy as well as for sketching the positions of bioenergy actors in debates. Thus, the following section presents some data on bioenergy in Germany. In Sect. 6.3.2 and 6.3.3, I explore the relationship of bioenergy and R&I in the German Bioeconomy Strategy.

6.3.1 The Socio-Energy Nexus in Germany's Transition Towards Renewable Energies

Legal regulation has led to a boom in the energy transition in Germany: with the implementation of the 1990 feed-in law (*Stromeinspeisegesetz*) and, notably, the 2000 Renewable Energies Act (EEG), the share of renewables in gross electricity consumption increased from 3.4% in 1990 to 37.8% in 2018, and the share of end consumption for heating rose from 2.1% to 13.9%, and in the transport sector from

0.1% to 5.6% during the same period (see BMWi—Federal Ministry for Economic Affairs and Energy 2019, pp. 4–5). Biomass-based energy produced by farmers and operators of bioenergy plants using cultivated biomass (including wood), cow manure, and other waste and residues plays a part in this renewable energy scenario. The Renewable Energy Agency refers to bioenergy as the "all-rounder"[4] amongst the renewables, although its share in the energy mix is decreasing in Germany: in 2019, bioelectricity, which is mostly produced in biogas plants in Germany, accounted for about 8.7% of total gross electricity consumption (UBA—Federal Environment Agency 2020). In the heating sector, biomass-based energy (notably from wood) makes up for 86% and is the most important renewable in the sector. Renewable fuels (biodiesel, bioethanol, biomethane) have been constant at around 5.6%, although the share of electricity-based fuels has increased recently (see ibid.; FNR—Fachagentur nachwachsende Rohstoffe 2019). Changes in the EEG (2012 and 2014) restricted the yearly extension of bioelectricity to 0.1 Gigawatt (GW; for comparison: Photovoltaic: 2.5 GW), to the discontent of many bioenergy producers (Haas 2017, pp. 64–210). Most of this power was produced from waste and residues. The growth of bioenergy production and consumption has been accompanied by controversy about monocultures, rising land prices and competition between biomass production for food, fodder, and energy (see 6.3.2). This has led to contestation over sustainability policies for bioenergy and, amongst others, calls for investment in bioenergy produced using waste, residues, and microorganisms.[5]

Concerning actor constellations and ownership structures of the transitioning energy system, the German *Energiewende* was born out of the struggles of social movements against nuclear power and for environmental protection as well as of efforts by pioneering citizens and small enterprises that invested in and experimented with renewable technologies. A large percentage of renewable energy continues to be produced by

---

[4] See, https://www.unendlich-viel-energie.de/erneuerbare-energie/bioenergie. Accessed 3 Feb 2020.

[5] See, https://biooekonomie.de/nachrichten/industrie-setzt-auf-bioenergie-der-zweiten-generation. Accessed 12 April 2020.

small-scale producers, be it private farmers, building owners, or citizen-based energy projects (for a comprehensive overview, see Kahla 2018). Bioenergy is part of the product range provided by energy cooperatives, mostly for consumption in the (rural) region in which it is produced (this especially applies to electricity and heating, but also to fuel, e.g. for agricultural vehicles). Bioenergy villages and regions, which are locations where citizens consume energy generated from local resources in plants and facilities owned by cooperatives or by private residents, have been promoted by governments to enlarge income opportunities for agricultural production and forestry, advance rural livelihoods and support acceptance of the energy transition, and ensure that it becomes anchored within society (see Hirschl et al. 2010; Kunze 2012). Recently, local heating networks based on renewables have begun to gain in importance, since the heating sector now accounts for around 50% of the energy demand (Beer et al. 2018, pp. 74–76).

In sum, bioenergy in Germany ranges from (large-)scale and ecologically problematic monocultures to the use of waste and residues, and from large biogas plants to small facilities that particularly generate electricity and heat, and which are often based on models of citizen participation. This social dimension of bioenergy infrastructure is important for debates about the decentralization and democratic design of the energy system as well as about the bioeconomy's aim to use biomass for it illustrates how transformation strategies such as the bioeconomy can be anchored in society. I now turn to bioenergy epistemics in the German bioeconomy, arguing that it neglects these aspects.

6.3.2 Bioenergy Epistemics: Funding of Knowledge Production and Narratives

In Germany, energy research includes energy conversion, storage systems and energy efficiency, institutional energy, and nuclear security. Funding of applied research into (bio)energy applications for specific projects is undertaken under the lead of the BMWi, whereas basic research is the responsibility of the BMBF (BMWi 2018a, b). In the National Research Strategy Bioeconomy 2030 (NRS) (BMBF 2010), which was launched by the BMBF, bioenergy is the least funded "field of action" (FoA). Until 2016, FoA 5 (Developing biomass-based energy carriers) received just 2.6% of funds (22 million euros; in contrast, research on renewable resources for industry received 23.7% or 204.6 million euros) (Fraunhofer ISI 2017, pp. 3–5).[6] Bioenergy-related research has been undertaken in other areas under the responsibility of the BMEL, the BMBF, and the BMWi (ibid.). However, it is irritating that although documents relating to the Bioeconomy Strategy list projects funded by these ministries, other documents produced by the BMWi on bioenergy research pay little or no attention to the Bioeconomy Strategy (e.g. BMWi 2011, 2018a, b).[7] The minor role of bioenergy played here is a reflection of the decreasing importance of the one-to-one substitution of fossil energy resources by biomass-based ones within bioeconomy strategies (Vivien et al. 2019, pp. 192–193). Instead, the focus is increasingly on cascade use and by-products (e.g. BMEL 2016, pp. 75–76).

Experts suggest two reasons for this: first, the biotechnology sector has predominantly pushed the bioeconomy (Grefe 2016) and the German as well as the EU strategy (EU 2007) are successors of biotechnology policies (Fraunhofer ISI 2017, p. 1; Lühmann 2020), with the latter having provided the impulse for the former (Kleinschmit et al. 2017, p. 5). Like other strategies focusing on the advanced use of biomass and the involvement of different economic sectors (Vivien et al. 2019), the German strategy stands "in the tradition of past expectations of biotechnology" and expands "the promises of economic growth to traditional sectors of the bioeconomy" (Meyer 2017, p. 7). Bioenergy has the lowest added value, and despite public funding and incentives, research on and the production of bioenergy exist under market conditions; hence, bioenergy has to compete with fossil fuels and the well-established production chains associated with petrochemicals, which are obstacles to bio-based products with higher value generation (ibid., pp. 17–18). Further, bioenergy also faces competition from various renewable energy options and from different sectors of the bioeconomy that are dependent on biomass

---

[6] 36 funding measures were accepted under the umbrella of the BMBF alone, and 1,800 projects were funded between 2009 and 2016 with 876 million euros (Fraunhofer ISI 2017, p. 1).
[7] Further research should engage with these deviating perceptions on the importance of the Bioeconomy Strategy in the state apparatus, possibly reflecting competing resort responsibilities.

(ibid., p. 23). The role of bioenergy in the future energy system will depend on technological innovation that enables multi-purpose and upgraded agriculture and forestry and, therefore, higher value generation, as this provides new fields for capital accumulation. Research is thus focused on the application maturity of products and innovative technologies, and on its capital- and research-intensive forms, in particular. This applies not only to a "technic-centred understanding of innovations" (ibid., p. 9; Backhouse et al. 2017) but also to the future trajectory of the energy transition. In the case of bioenergy, innovation exists and research is funded in areas such as the flexibilization of biogas plants, as well as into biomethane production and cogeneration plants (see DUH—Deutsche Umwelthilfe 2018), and, as part of the Bioeconomy Strategy, into the use of frugal energy plants, waste, and residues (e.g. BMEL 2016, pp. 59, 68). However, when describing biogas, an expert argued that biogas was an "old hat" that provided no "new technology, no new approach, and thus it is not of real political value" (Institute for Biogas, Interview no. 1, own translation) in contrast to, e.g. power-to-gas-plants. Hence, in this view, other energy sources can be politically promoted much better as innovative enough to meet the energy demands of a post-fossil Germany.

This character of the bioeconomy is driven by the focus on new technologies which promises to secure both the status quo of resource consumption and environmental protection (see Birch et al. 2010). It is reflected in the funding of projects that use algae for bio-kerosene for aviation and straw for biofuel (in cooperation with leading stock companies such as Airbus Group, OMV, and Clariant AG). The Recommendations of the German Bioeconomy Council for energy in the bioeconomy do not even mention the word "biomass" in the relevant paragraph on the "Conversion and storage of solar energy, hybrid energy systems" (German Bioeconomy Council 2016, p. 16), which, instead, focuses on artificial photosynthesis and solar-based energy production. The accompanying narrative to this focus on high-technology innovation is based on perspectives linked to Germany as an industrial location for technology leadership and international competitiveness (Fraunhofer ISI 2017, p. 1; BMEL 2016, p. 68; German Bioeconomy Council 2016,

p. 11); it also reflects Germany's political economy as an export-driven regime of accumulation (see Haas 2017, pp. 146–154).

Secondly, experts from the bioenergy sector assume that the food-or-fuel-debate that has taken place during the past two decades has led (amongst other factors concerning the economic efficiency of bioenergy) to legal changes in areas including the regulation of the use of renewable cultivated biomass such as corn (Beer et al. 2018, p. 75). This debate revolves around the argument that the use of agricultural or forest lands for the cultivation of energy crops triggers competition between food, feed, or fodder production and, frequently, leads to negative socio-ecological impacts (see Evia Bertullo 2018; Lewandowski 2015). Press articles, blog contributions, as well as recent qualitative interviews by the author of this contribution reveal that actors in the bioenergy sector such as biogas producers, employees of state agencies, and consultants consider this critique of bioenergy as unfair as it conceals the technological and ecological progress that has been achieved in the sector. Although comprehensive studies into the influence of this debate on bioeconomy policies are a lacuna, Meyer suggests that the controversies surrounding first generation biofuels could be germane to the "aspired transformation towards a bio-based industry" (Meyer 2017, p. 23). If it is mentioned in the bioeconomy context, bioenergy research is framed as sustainable, rural development and a commitment to food-first, although the latter has yet to be proven (ibid., p. 23; see also Bringezu et al. 2020).

6.3.3 Bioenergy Justice: R&I Innovations and Societal Participation

Civil society actors have concentrated on the development of the exclusive research agendas within the Bioeconomy Strategies from the start. They argue that a reliance on old and "antiquated" (BUND, Interview no. 2, own translation; also Meyer 2017, pp. 16–17) structures and networks of interest intermediation between state agencies, private economic actors, and scientists in Germany (see Brandl 2018) has frequently led to the exclusion of civil society actors (e.g. Grefe 2016; Zivilgesellschaftliche Plattform Forschungswende 2017). In the wake of this criticism, the composition of the expert German Bioeconomy Council, which mostly consists of scientific experts from the biotechnology and chemical industry, slightly changed and stakeholder workshops have increasingly included critical NGOs and experts. However, the Council continues to be dominated by (non-organic) agricultural and consumer associations and the biotech and chemistry industry. Further, although a significant number of environmental NGOs and organic farmers' associations have participated in events run by government institutions that explicitly address issues like an environmental friendly agriculture or biodiversity protection,[8] "dialogue" is limited to these formats; as such, no structured funding exists for critical NGOs to engage with complex bioeconomy-related issues or to set up R&I agendas and compete politically with the private sector (see Zivilgesellschaftliches Aktionsforum Bioökonomie 2019).

The inclusion of critical or alternative research in the bioeconomy predominantly concerns agriculture, with a notable concentration on agroecological practices (Meyer 2017, p. 22; see Levidow et al. 2012). However, in line with the minor involvement of NGOs and other civil society-based associations, collective knowledge about energy issues and practices of societal control of and participation in renewable energy projects by energy cooperatives and energy justice activists is scarcely included in the bioenergy-related research that is conducted within the bioeconomy context. Only one event listed in past funding activities on bioeconomy stakeholder meetings explicitly addresses some of these actor groups: the "Congress Bioenergy Villages" (Kongress Bioenergiedörfer) that took place in March 2014 (Bundesregierung 2017, p. 12). Although similar events are financed by other state programmes and agencies, this demonstrates that the experiences of these actors do not shape the bioeconomy agenda.

In the light of the character of bioenergy epistemics in the bioeconomy, one could argue that the promise that in the case of citizen-based bioenergy production R&I will lead to the same extent

---

[8]Unions did not participate in either of these events (Bundesregierung 2017, p. 10–15), although the dominant form of the bioeconomy and the specific mode of value creation will reproduce and create different kinds of jobs, varying in terms of work relations, work locations (urban, rural), or requested education (Braun and Brandl 2016; see Lorenzen in this volume).

of economic growth as other forms of research on bioeconomy-technoscience have not been kept. These models are provided with some funding as part of BMWi's energy policies, with emphasis on the digitalization of energy production, distribution, and consumption. In addition, in electricity and heating, which is the concern of federal, regional, and local policy-making, practices of decentralized structures exist that also receive political support (Beer et al. 2018, pp. 68–70, 76–77). An energy justice perspective could apply these experiences of the decentralized control of important utilities (energy) when dealing with current and future developments related to more advanced biomass-based products.

However, R&I related to bio-based chemicals, products, and energy technology is, as Frickel et al. (2010, p. 467) state when referring to technoscience in general, "increasingly complex, technology-laden, and expensive". Studies of bioenergy cooperatives as well as recent interviews with experts in the field show that the complexity of technology, administration, and organization of (cooperative) bioenergy production facilities leads to the automatic exclusion of cooperative members or farmers who lack the time and knowledge to hold pace with technological and regulatory advancements (see Backhouse et al. 2020). Although the German Bioeconomy Strategy is celebrating its tenth anniversary, qualitative research for this chapter shows that experts and practitioners from the bioenergy sector have, at best, a vague understanding of the bioeconomy pushed by the federal government. Interviews showed that, the sector is focused on the expiration of subsidies for biogas plants, wood carburettors and wood chip heating, as well as unstable energy transition regulations and environmental laws like the recent decree on liquid manure, which implies investment for stock farmers. The question of how further R&I in the bioenergy-related field of the bioeconomy—innovation-driven, technology-laden—reproduces or advances this tendency, or if broader sectors of society are included in distributional, recognition-based, and procedural aspects of future (bio)energy questions, remains open.

6.4 Conclusion

In this chapter, I examined inclusions and exclusions in bioenergy-related research funding activities in the German Federal Government's first Bioeconomy Strategy against the backdrop of theoretical concepts that stress power relations and inequalities in energy transition politics as well as knowledge production. I adopted the perspective of energy justice research and focused on knowledge production in bioenergy. Moreover, I particularly focused on the role of decentralized energy production practices, since it is these social forms that enhance inclusion and partly enabled the development of the German energy transition. I asked how and to what extent bioenergy is related to the production of R&I in the Bioeconomy Strategy and the kind of knowledge about energy production that is supported by funding strategies; the aim was to assess the prospects of the social side of bioenergy in future socio-energy systems as well as in the bioeconomy.

Findings show that unequal energy epistemics are reflected in the fissure between technology-laden research and existing practices of bioenergy production. Bioeconomy research into bioenergy reveals the constricted perception of innovation and the assumption that R&I has to engender economic growth. Existing practices of different actors in the bioenergy field tend not to be taken into account by those who design the bioeconomy agenda. A structured exchange is still lacking; this applies to the Bioeconomy Strategy in general, as well as to bioeconomy-related bioenergy policies specifically. The different social experiences that have been made with renewable (bio)energy in Germany provide an opportunity not only to broaden the spectrum of stakeholder discussions and funding schemes but also to engender structural change that could set the framework with which to reshape the character and goals of research within the bioeconomy, which, until now, has been rather exclusive and technology-laden. Given the contention surrounding planned renewable energy infrastructures, the inclusion of more actors and a broader perspective on innovation and its use for society would open the door for discussions about just transitions—this applies to both socio-energy systems and the resource basis of current economies. The extent and regard to which civil society-based bioenergy production models will be included or include themselves in (discussions about) bioenergy-related bioeconomy policies and whether the idea of decentralized, democratized energy systems can be integrated or combined with an increasingly technology-laden and digitalized production, distribution, and consumption of energy are of further interest.

The draft and final version of the new Bioeconomy Strategy, which passed cabinet in January 2020 (BMBF and BMEL 2019, 2020), resonates the need for research into a broader spectrum of bioeconomic practices and knowledge, including studies of environmental or development impacts, and resource competition. The way in which this will be realized deserves to be a focus of research in the years to come.

List of Interviews quoted

Expert interview no.	Institution; Organization	Date and place
Expert interview no. 1	Institute for Biogas, Waste Management and Energy	16 January 2020, Weimar
Expert interview no. 2	NABU (Nature And Biodiversity Conservation Union)	17 October 2019, Berlin

References

Arancibia, F., & Motta, R. (2018). Undone Science and Counter-Expertise: Fighting for Justice in an Argentine Community Contaminated by Pesticides. *Science as Culture, 28*(3), 277–302.

Avila, S. (2018). Environmental Justice and the Expanding Geography of Wind Power Conflicts. *Sustainability Science, 13*(3), 599–616.

Backhouse, M., Lorenzen, K., Lühmann, M., Puder, J., Rodríguez, F., & Tittor, A. (2017). Bioökonomie-Strategien im Vergleich. Gemeinsamkeiten, Widersprüche und Leerstellen. *Working Paper 1, Bioeconomy & Inequalities*, Jena. https://www.bioinequalities.uni-jena.de/sozbemedia/neu/2017-09-28+workingpaper+1.pdf. Accessed 14 Nov 2019.

Backhouse, M., Büttner, M., Greifenberg, D., Herdlitschka, T., Lehmann, R., Schaller, E., et al. (2020). Erneuerbare Energien von unten? Perspektiven aus der Praxis auf dezentrale Energiesysteme. *Working Paper 14, Bioeconomy & Inequalities*, Jena. https://www.bioinequalities.uni-jena.de/sozbemedia/wp/workingpaper14.pdf. Accessed 10 June 2020.

Becker, S., & Naumann, M. (2017). Energy Democracy: Mapping the Debate on Energy Alternatives. *Geography Compass, 11*(8), 1–13.

Bentsen, N.S., Larsen, S., & Stupak, I. (2019). Sustainability Governance of the Danish Bioeconomy—The Case of Bioenergy and Biomaterials from Agriculture. *Energy, Sustainability and Society, 9*(40).

Beer, K., Böcher, M., Bollmann, A., Töller, A.E., & Vogelpohl, T. (2018). Arbeitsbericht 1. Fallauswahl und Übersichtsanalysen. Verbundprojekt Politische Prozesse der Bioökonomie zwischen Ökonomie und Ökologie. http://www.bio-oekopoli.de/bio-oekopoli/download/arbeitspapier_1.pdf. Accessed 14 Nov 2019.

Birch, K., Levidow, L., & Papaioannou, T. (2010). Sustainable Capital? The Neoliberalization of Nature and Knowledge in the European "Knowledge-Based Bio-economy". *Sustainability, 2*, 2898–2918.

BMBF (2010). *National Research Strategy BioEconomy 2030. Our Route towards a biobased economy*. Bonn. https://www.pflanzenforschung.de/application/files/4415/7355/9025/German_bioeconomy_Strategy_2030.pdf. Accessed 5 Nov 2020.

BMBF (2014). *Wegweiser Bioökonomie. Forschung für biobasiertes und nachhaltiges Wirtschaftswachstum*. Berlin. https://biooekonomie.de/sites/default/files/publications/wegweiser-biooekonomiepropertypdfbereichbiooekospracherderwbtrue.pdf. Accessed 5 Nov 2020.

BMBF & BMEL (2014). *Bioeconomy in Germany. Opportunities for a Bio-based and Sustainable Future*. Bonn, Berlin. https://www.bmbf.de/upload_filestore/pub/Biooekonomie_in_Deutschland_Eng.pdf. Accessed Nov 23 2020.

BMBF & BMEL (2019). *Nationale Bioökonomiestrategie*. 2 July 2019 ENTWURF. https://www.raiffeisen.de/sites/default/files/2019-07/2019-07-02_Nationale%20Bio%C3%B6konomiestrategie.pdf. Accessed 30 Sep 2019.

BMBF & BMEL (2020). *National Bioeconomy Strategy*. Berlin. https://www.bmbf.de/upload_filestore/pub/BMBF_Nationale_Biooekonomiestrategie_Langfassung_eng.pdf. Accessed 19 Oct 2020.

BMEL (2014). *Nationale Politikstrategie Bioökonomie*. Berlin. https://biooekonomie.de/sites/default/files/publications/npsb_0.pdf. Accessed 5 Nov 2020.

BMEL (2016). *Fortschrittsbericht zur Nationalen Politikstrategie Bioökonomie*. Berlin. https://www.bmel.de/SharedDocs/Downloads/DE/Broschueren/For

tschrittsbericht-Biooekonomie.pdf?__blob=publicationFile&v=2. Accessed 5 Nov 2020.

BMWi (2011). Forschung für eine umweltschonende, zuverlässige und bezahlbare Energieversorgung. Das 6. Energieforschungsprogramm der Bundesregierung. Berlin. https://www.bmwi.de/Redaktion/DE/Publikationen/Energie/6-energieforschungsprogramm-der-bundesregierung.pdf?__blob=publicationFile&v=12. Accessed 5 Nov 2020.

BMWi (2018a). *2018 Federal Government Report on Energy Research. Funding Research for the Energy Transition*. Berlin. https://www.bmwi.de/Redaktion/EN/Publikationen/Energie/bundesbericht-energieforschung-2018.pdf?__blob=publicationFile&v=5. Accessed 5 Nov 2020.

BMWi (2018b). *Innovationen für die Energiewende. 7. Energieforschungsrahmenprogramm der Bundesregierung*. Berlin. https://www.bmwi.de/Redaktion/DE/Publikationen/Energie/7-energieforschungsprogramm-der-bundesregierung.pdf?__blob=publicationFile&v=4. Accessed 5 Nov 2020.

BMWi (2019). *Zeitreihen zur Entwicklung der erneuerbaren Energien in Deutschland*. Berlin. https://www.erneuerbare-energien.de/EE/Redaktion/DE/Downloads/zeitreihen-zur-entwicklung-der-erneuerbaren-energien-in-deutschland-1990-2019.pdf;jsessionid=89DA6B4D4C703767C26DF37599690377?__blob=publicationFile&v=26. Accessed 18 Jan 2020.

Brandl, B. (2018). *Wissenschaft, Technologieentwicklung und die Spielarten des Kapitalismus. Analyse der Entwicklung von Saatgut in USA und Deutschland*. Wiesbaden: Springer VS.

Braun, V., & Brandl, B. (2016). Von der Kommodifizierung zur Refeudalisierung? Wertschöpfung in der Bioökonomie. Beitrag zur Veranstaltung ‚Bioökonomie. Grenzen des Wachstums oder Füllhorn Natur?' Sektion Land- und Agrarsoziologie der Deutschen Gesellschaft für Soziologie.

Bringezu, S., Banse, M., Ahmann, L., Bezama, A., Billig, E., Bischof, R., et al. (2020). Pilotbericht zum Monitoring der deutschen Bioökonomie. Universität Kassel, Center for Environmental Systems Research (CESR). https://kobra.uni-kassel.de/handle/123456789/11591. Accessed 8 Aug 2020.

Bundesregierung (2017, April 21). Antwort der Bundesregierung auf die Kleine Anfrage der Abgeordneten Harald Ebner, Kai Gehring, Friedrich Ostendorff, weiterer Abgeordneter und der Fraktion Bündnis 90/DIE GRÜNEN: Weiterentwicklung der „Nationalen Forschungsstrategie Bioökonomie 2030". Deutscher Bundestag, 18. Wahlperiode, Drucksache 18/12024. http://dipbt.bundestag.de/doc/btd/18/120/1812024.pdf. Accessed 5 Nov 2020.

DUH (2018). Protokoll Netzwerktreffen Bioökonomie, 28 Nov 2018, Berlin. Nicht veröffentlichtes Dokument.

Elmhirst, R., Siscawati, M., Sijapati Basnett, B., & Ekowati, D. (2017). Gender and Generation in Engagements with Oil Palm in East Kalimantan, Indonesia: Insights from Feminist Political Ecology. *The Journal of Peasant Studies, 44*(6), 1135–1157.

EU (2007). *En Route to the Knowledge-Based Bio-Economy*. Brussels. https://dechema.de/dechema_media/Downloads/Positionspapiere/Cologne_Paper.pdf. Accessed 5 Nov 2020.

Evia Bertullo, V. (2018). Saberes y Experiencias sobre la Exposición a Plaguicidas entre Mujeres que Residen en Contextos Agrícolas en Soriano, Uruguay. *Revista TRAMA, 9*(9), 13–35.

FNR (2019). Bioenergy in Germany. Facts and Figures 2019. http://www.fnr.de/fileadmin/allgemein/pdf/broschueren/broschuere_basisdaten_bioenergie_2018_engl_neu.pdf. Accessed 20 Jan 2020.

Fraunhofer ISI (2017). Evaluation der „Nationalen Forschungsstrategie BioÖkonomie 2030". Wirksamkeit der Initiativen des BMBF – Erfolg der geförderten Vorhaben – Empfehlungen zur strategischen Weiterentwicklung. https://www.isi.fraunhofer.de/content/dam/isi/dokumente/cct/2017/Evaluation_NFSB_Abschlussbericht.pdf. Accessed 14 April 2019.

Frickel, S., Gibbon, S., Howard, J., Kempner, J., Ottinger, G., & Hess, D. (2010). Undone Science: Charting Social Movement and Civil Society Challenges to Research Agenda Setting. *Science, Technology, & Human Values, 35*(4), 444–473.

German Bioeconomy Council (2016). *Recommendations of the German Bioeconomy Council. Further Development of the "National Research Strategy BioEconomy 2030"*. Berlin. https://www.biooekonomierat.de/fileadmin/Publikationen/Englisch/BOER_Empfehlungspapier_ENG_final.pdf. Accessed 23 Nov 2020.

Giurca, A. (2018). Unpacking the Network Discourse: Actors and Storylines in Germany's Wood-Based Bioeconomy. *Forest Policy and Economics, 110*.

Goven, J., & Pavone, V. (2014). The Bioeconomy as Political Project: A Polanyian Analysis. *Science, Technology, & Human Values, 40*(3), 302–337.

Görg, C. (2004). The Construction of Societal Relationships with Nature. *Poiesis & Praxis, 3*(1), 22–36.

Grefe, C. (2016). *Global Gardening. Bioökonomie – Neuer Raubbau oder Wirtschaftsform der Zukunft?* München: Verlag Antje Kunstmann.

Haas, T. (2017). *Die politische Ökonomie der Energiewende. Deutschland und Spanien im Kontext multipler Krisendynamiken in Europa.* Wiesbaden: Springer VS.

Healy, N., & Barry, J. (2017). Politicizing Energy Justice and Energy System Transitions: Fossil Fuel Divestment and a "Just Transition". *Energy Policy, 108,* 451–459.

Hess, D.J., & McKane, R.G. (2017). Renewable Energy Research and Development. A Political Economy Perspective. In D. Tyfield, R. Lave, S. Randalls & C. Thorpe (Eds.), *The Routledge Handbook of the Political Economy of Science* (pp. 275–288). London, New York: Routledge.

Hirschl, B., Aretz, A., Prahl, A., Böther, T., Heinbach, K., Pick, D., et al. (2010). Kommunale Wertschöpfung durch erneuerbare Energien. Schriftenreihe des IÖW 196/10. https://www.ioew.de/uploads/tx_uki oewdb/IOEW_SR_196_Kommunale_Wertsch%C3%B6pfung_durch_E rneuerbare_Energien.pdf. Accessed 17 June 2020.

Huber, M. (2015). Energy and Social Power. From Political Ecology to the Ecology of Politics. In T. Perreault, G. Bridge & J. McCarthy (Eds.), *The Routledge Handbook of Political Ecology* (pp. 481–492). Abingdon, New York: Routledge.

Jenkins, K., McCauley, D., Heffron, R., Stephan, H., & Rehner, R. (2016). Energy Justice: A Conceptual Review. *Energy Research & Social Science, 11,* 174–182.

Kahla, F. (2018). Das Phänomen Bürgerenergie in Deutschland. Eine betriebswirtschaftliche Analyse von Bürgergesellschaften im Bereich der Erneuerbaren Energien-Produktion. Lüneburg: Leuphana Universität Lüneburg. https://pub-data.leuphana.de/frontdoor/deliver/index/docId/848/file/Dissertation_Kahla.pdf. Accessed 17 June 2020.

Kleinschmit, D., Arts, B., Giurca, A., Mustalahti, I., Sergent, A., & Pülzl, H. (2017). Environmental Concerns in Political Bioeconomy Discourses. *International Forestry Review, 19*(1), 1–15.

Kunze, C. (2012). *Soziologie der Energiewende: Erneuerbare Energien und die Transition des ländlichen Raums.* Stuttgart: ibidem-Verlag.

Levidow, L., Birch, K., & Papaioannou, T. (2012). Divergent Paradigms of European Agro-Food Innovation: The Knowledge-Based Bio-Economy (KBBE) as an R&D Agenda. *Science, Technology, & Human Values, 38*(1), 94–125.

Lewandowski, I. (2015). Securing a Sustainable Biomass Supply in a Growing Bioeconomy. *Global Food Security, 6,* 34–42.

Lohmann, L., & Hildyard, N. (2014). Energy, Work and Finance. The Corner House. http://www.thecornerhouse.org.uk/sites/thecornerhouse.org.uk/files/EnergyWorkFinance%20%282.57MB%29.pdf. Accessed 10 Oct 2019.

Lühmann, M. (2020). Whose European Bioeconomy? Relations of Forces in the Shaping of an Updated EU Bioeconomy Strategy. *Environmental Development*, 35.

Malm, A. (2016). *Fossil Capital. The Rise of Steam Power and the Roots of Global Warming*. New York, London: Verso.

Marshall, J. (2014). Solar Energy. Springtime for the Artificial Leaf. *Nature*. https://www.nature.com/news/solar-energy-springtime-for-the-artificial-leaf-1.15341. Accessed 15 May 2019.

Meyer, R. (2017). Bioeconomy Strategies: Contexts, Visions, Guiding Implementation Principles and Resulting Debates. *Sustainability*, 9(6), 1031.

Miller, C.A., Iles, A., & Jones, C.F. (2013). The Social Dimension of Energy Transitions. *Science as Culture*, 22(2), 135–148.

Mitchell, T. (2009). Carbon Democracy. *Economy and Society*, 38(3), 399–432.

Morris, C., & Jungjohann, A. (2016). *Energy Democracy. Germany's Energiewende to Renewables*. Houndmills, Basingstoke: Palgrave Macmillan.

Priefer, C., Jörissen, J., & Frör, O. (2017). Pathways to Shape the Bioeconomy. *Resources*, 6(1), 10.

Radtke, J. (2016). *Bürgerenergie in Deutschland. Partizipation zwischen Gemeinwohl und Rendite*. Wiesbaden: Springer VS.

Strzalka, R., Schneider, D., & Eicker, U. (2017). Current Status of Bioenergy Technologies in Germany. *Renewable and Sustainable Energy Reviews*, 72, 801–820.

Szarka, N., Eichhorn, M., Kittler, R., Bezama, A., & Thrän, D. (2017). Interpreting Long-Term Energy Scenarios and the Role of Bioenergy in Germany. *Renewable and Sustainable Energy Reviews*, 68, 1222–1233.

Szulecki, K. (2018). Conceptualizing Energy Democracy. *Environmental Politics*, 27(1), 21–41.

Tyfield, D., Lave, R., Randalls, S., & Thorpe, C. (2017). Introduction: Beyond Crisis in the Knowledge Economy. In D. Tyfield, R. Lave, S. Randalls & C. Thorpe (Eds.), *The Routledge Handbook of the Political Economy of Science* (pp. 1–18). London, New York: Routledge.

UBA (2020). Erneuerbare Energien in Deutschland. Daten zur Entwicklung im Jahr 2019. https://www.umweltbundesamt.de/sites/default/files/medien/1410/publikationen/2020-04-03_hgp-ee-in-zahlen_bf.pdf. Accessed 3 May 2020.

Vivien, F.-D., Nieddu, M., Befort, N., Debref, R., & Giampietro, M. (2019). The Hijacking of the Bioeconomy. *Ecological Economics, 159*, 189–197.

Weis, L., Becker, S., & Naumann, M. (2015). Energiedemokratie. Grundlage und Perspektive einer kritischen Energieforschung. Rosa-Luxemburg-Stiftung STUDIEN (01). https://www.rosalux.de/fileadmin/rls_uploads/pdfs/Studien/Studien_01-15_Energiedemokratie.pdf. Accessed 10 Oct 2019.

Wissen, M. (2016). Jenseits der carbon democracy. Zur Demokratisierung der gesellschaftlichen Naturverhältnisse. In A. Demirović (Ed.), *Transformation der Demokratie – demokratische Transformation* (pp. 48–66). Münster: Westfälisches Dampfboot.

Zivilgesellschaftliches Aktionsforum Bioökonomie (2019, July 19). Stellungnahme an die Bundesregierung zum Entwurf einer Nationalen Bioökonomiestrategie. https://denkhausbremen.de/wp-content/uploads/2019/07/Stellungnahme-Bio%C3%B6konomie.pdf. Accessed 14 Nov 2019.

Zivilgesellschaftliche Plattform Forschungswende (2017). Positionspapier zivilgesellschaftliche Plattform Forschungswende: Kernforderungen für das 7. Energieforschungsrahmenprogramm. https://www.energieforschung.de/lw_resource/datapool/systemfiles/elements/files/6FFE14A82C281782E0539A695E862C2E/current/document/5_FW_Positionspapier_Energie.pdf. Accessed 4 Oct 2019.

Open Access This chapter is licensed under the terms of the Creative Commons Attribution 4.0 International License (http://creativecommons.org/licenses/by/4.0/), which permits use, sharing, adaptation, distribution and reproduction in any medium or format, as long as you give appropriate credit to the original author(s) and the source, provide a link to the Creative Commons license and indicate if changes were made.

The images or other third party material in this chapter are included in the chapter's Creative Commons license, unless indicated otherwise in a credit line to the material. If material is not included in the chapter's Creative Commons license and your intended use is not permitted by statutory regulation or exceeds the permitted use, you will need to obtain permission directly from the copyright holder.

7

A Player Bigger Than Its Size: Finnish Bioeconomy and Forest Policy in the Era of Global Climate Politics

Tero Toivanen

7.1 Introduction

During the last few decades, the bioeconomy has become a key feature in framing the transition to a sustainable future. Transnational economic organisations and many countries have drawn up ambitious bioeconomy strategies. Recently, some governments in industrialised countries have adopted these strategies in order to strike a path that goes beyond the fossil economy (Meyer 2017). These strategies represent bioeconomy as a sustainable solution that mitigates climate change and other environmental problems as well as creating the next generation of sustainable products and fostering green economic growth. Thus, bioeconomy incorporates a remarkably wide set of ideas and economic activities under one inspiring and highly optimistic umbrella term (Birch 2006; Birch and Tyfield 2013; Bugge et al. 2016).

T. Toivanen (✉)
BIOS Research Unit, Helsinki, Finland
e-mail: tero.toivanen@helsinki.fi

Bioeconomy is particularly important in countries with a large forestry sector. Forests are one of the most promising resources for societies that are searching for ways to replace the fossil economy. Simultaneously, the flexibility of the concept enables the forest industry to reframe its traditional industrial operations in new, greener terms: in the blink of an eye, pulp factories become "biorefineries" or "bioproduct factories". Finland is a case in point. During the last decade, Finland has provided a platform for a successful *bioeconomic imaginary* (Goven and Pavone 2015). After decades of decline, this imaginary has significantly contributed to a development that has relocated the traditional forest industry at the heart of the Finnish national political economy. These bioeconomic developments have prepared the ground for a new *forest policy regime* (see Donner-Amnell et al. 2004; Kotilainen and Rytteri 2011), the bioeconomy regime. As a result, the forest bioeconomy emerged as an important, if not the most important, policy of the Finnish centre-right government that was in power between 2015 and 2019.

However, a novel scientific factor now poses a challenge to the image of sustainability that shrouds the forest bioeconomy. The rapid climate mitigation targets that were put in place in the wake of the Paris Climate Agreement not only entail radically slashing emissions from fossil fuel but also removing carbon dioxide from the atmosphere. The world is to achieve carbon neutrality around mid-century, with developed countries expected to reach carbon neutrality a lot earlier. Finland is to be carbon neutral as close as possible to 2030 (FCCP—Finnish Climate Change Panel 2018).

The world's forests have a crucial role to play in this global and national "Herculean task" (Rockström et al. 2017): as part of land-based ecosystems, forests are the only functioning carbon sinks that can increasingly remove carbon from the atmosphere. Thus, the best way of achieving rapid climate mitigation is to develop global carbon sinks by stopping deforestation, significantly reducing harvesting and implementing reforestation projects. Obviously, if the role of forests is framed within these terms, the politics of carbon sinks at the global level have enormous political, economic and social importance in countries with large forestry sectors, such as Finland.

The issue of carbon sinks burst onto the stage as part of the Finnish public debate soon after the bioeconomic imaginary had successfully repositioned the forest sector at the core of national political economy. In a short period of time, the world of Finnish forestry seemed to have been turned upside down: the bioeconomy plans, which had hitherto sailed along in fair winds, suddenly came to be questioned. Finnish climate and forest researchers delivered an unpleasant message, which was underscored by the tightening of EU climate policies: Finland's bioeconomy strategies—which were based on increasing the rate at which forests were harvested and, thus, reducing the size of forest carbon sinks—would certainly not be able to mitigate climate change in the time available.

In this paper, I study how the results of scientific research on the role of forests in climate change mitigation challenged the Finnish bioeconomy regime. I analyse the key developments of a four-year debate from 2015 to 2019 on forest carbon sinks with a special focus on how actors closely related to the forestry sector reacted to the messages brought up by researchers. I rely on frame analysis, which has been widely practised in media studies, to understand how journalism creates and reinforces certain ideas in society (e.g. Entman 2007; Harjuniemi 2019). Framing collects certain aspects of a perceived reality and reformulates them as a narrative that promotes a particular interpretation. Frames introduce and enhance the importance of certain ideas in public discussion and activate "schemas" that encourage target audiences to think, feel, discuss and decide in a particular way (Entman 2007, pp. 164–165). In the case of Finnish bioeconomy, framing the public debate has significantly contributed to forming and legitimising historical forest policy regimes (see also Peltomaa 2018).

Finland represents an important case study in global climate and bioeconomy politics. Its globally influential forest industry means that it can obtain a greater role in global climate politics than its small size might suggest and, thus, it can be considered as an influential *small and medium-sized power* (Eloranta et al. 2018) in global political economy. The role of forests in climate change has been debated widely in Finland. This makes the Finnish case interesting in an international context: How has a novel scientific message and the tightening of EU climate regulation challenged the existing forest bioeconomy regime? How did the regime respond and how has it attempted to defend its interests and power?

This paper proceeds as follows. In the first section, I analyse the special features of the Finnish forest bioeconomy regime. In the second section, I describe in detail how the issues of forest carbon sinks and new EU climate regulation have challenged the bioeconomy regime. In the third section, I offer an analysis of the three-phase development that occurred in Finnish public debates about carbon sinks between 2015 and 2019. In the last section, I discuss the international political importance of the Finnish bioeconomy debate.

7.2 Finnish Bioeconomy as a Forest Policy Regime

Forestry has had an enormous impact on the history of Finland. The turns in the political economy of forestry have been closely related to the transformations of society as a whole. Previous research has analysed Finnish forestry in the context of historical forest policy regimes (Donner-Amnell et al. 2004; Kotilainen and Rytteri 2011). Historically, forest policy regimes have consisted of long-term, quasi-permanent, social, political, economic and cultural arrangements that underlie governmental actions (Kotilainen and Rytteri 2011). Regimes have changed over time: from the nineteenth-century pre-industrial regime and the industrial regime during the two world wars to a regime that from 1970s onwards has incorporated some aspects of environmental sustainability (ibid.; Kröger and Raitio 2017).

Despite these historical transformations, some things have remained the same. The symbiosis between private forest owners and the forest industry has created the social, political and economic basis for the long-term development of Finnish forestry. Whereas forest industry has been responsible for production, private forest owners have taken care of planting, growing and marketing wood. The social power of both actors has been enforced through the establishment of central associations: the Finnish Forest Industries and the Central Union of Agricultural Producers and Forest Owners (MTK). Furthermore, state policies have been harnessed in multiple ways to support the industry by organising funding, investing in infrastructure, drawing up trade policies and encouraging applied scientific forest research (Siiskonen 2007; Kröger and Raitio 2017).

Finnish bioeconomy is so fundamentally connected to the utilisation of the country's forest resources that Finnish bioeconomy *is* forest bioeconomy. As such, bioeconomy in Finland marks a potential beginning for a new forest policy regime. One promising way to analyse the material development of the bioeconomy is the opposition between an *expansion frame* (which means that an industrial regime, despite all of the green rhetoric and policies, continues to organise production in traditional extractivist terms) and a *transformation frame* (policies that set in motion a sector-wide low-carbon, sustainable transition).

Finnish bioeconomy emerged at a particular historical moment. On the thresholds of the 2008 global economic crisis, Finland experienced a twofold industrial setback. First, and this already applied before the financial crisis, the traditional chemical forest industry, the long-time core of the export-led national economy, was facing a downturn. Second, at the end of the 2000s, a successful Finnish high-tech sector came tumbling down when its cornerstone, the mobile phone company Nokia, ran into deep problems and shut down its landmark mobile device division.

In these historical conditions, the idea of bioeconomy started to gain attraction. The six-party coalition government that was in power between 2011 and 2015 was the first to mention the idea of bioeconomy and did so in its 2011 manifesto. In 2014, the first official bioeconomy strategy for Finland was published, and the centre-right government that was in office between 2015 and 2019 eventually adopted bioeconomy as the core of its approach. As a result, attention in Finnish political economy shifted from the promise of a network society to the promotion of a deeper use of Finland's natural resources. This led to the introduction of a new techno-economic framework, the bioeconomy regime, aimed at industrial renewal and which combines the traditional forestry sector with the promise of innovations and bioproducts.

Recently, Ahlqvist and Sirviö (2019) have argued that settling the tension between urban and rural areas constitutes a material condition for a successful bioeconomy regime. The industrial restructuring that took place during and after the 2000s hit the Finnish periphery hardest. To solve the problems of rural areas, the advocates of the bioeconomy promised "new economic dynamics to emerge throughout the state space, fostered by new investment projects and state subsidies designed to update infrastructures in the peripheral regions" (ibid., p. 403). Simultaneously, the bioeconomic imaginary also appealed to the advocates of urban-led development: whereas the countryside would continue to play the role of resource periphery, the high-tech side of the bioeconomy fit well into the high-skilled and educated imaginary of the urban bourgeoisie. In addition, when the bioeconomy initiative also promises solutions to climate change, this leads to a potential political compromise in which "everyone wins".

The implementation of bioeconomy strategies depends on the election of a supportive government. Finnish bioeconomy has always been a project of the Centre Party, a party with its electoral base in rural areas. An interesting anecdote associated with Finnish politics is the fact that bioeconomy is strongly associated with the former leader of the Centre Party, Juha Sipilä, who was the prime minister from 2015 to 2019. Before the 2015 parliamentary elections, Finnish media was enthralled by this successful businessman who had jumped into politics. The future prime minister drove around the rural periphery of North Finland with his wood-burning carbon monoxide car and promoted bioeconomy as a key to a sustainable future in Finland. Thus, the urban-rural contradiction was also settled in this political character.

In 2015, Finnish bioeconomy finally had its moment when the Centre Party and the right-wing National Coalition party, which is associated with the urban bourgeoisie, formed a government. The election of the new centre-right government provided the Finnish forest sector reason for celebration after decades of uncertainty. The positive atmosphere culminated in the decision to build the Metsä Group's Äänekoski "bioproduct factory", the biggest investment in the history of Finnish Forest Industry. The factory was strongly supported by a wide political spectrum. Bioeconomy was booming, and the new centre-right government declared bioeconomy as its most important (by net monetary investment) priority project.

Nevertheless, the Finnish bioeconomic imaginary has also faced criticism. Before the negative effects of these bioeconomy plans on the climate were fully understood, the left-wing parties, the Social Democrats, the Left Alliance and the Greens were sceptical about the bioeconomy. Furthermore, environmental NGOs criticised the possible negative impact (e.g. loss of forest biodiversity) associated with bioeconomy (see, e.g., FANC—The Finnish Association for Nature Conservation 2014). Thus, critical voices identified forms of "green washing" in the bioeconomy discourse. Criticism has also been directed at the fact that the majority of the bioeconomy (in terms of volume) remains in traditional industrial products, namely paper and pulp—a fact that would support the continuity of the expansion frame over any supposed move towards a transformation frame.

Another important matter that defines the Finnish bioeconomy regime is forest bioenergy. Forest bioenergy composes one quarter of the total energy produced in Finland. In the renewable energy sector, forest-based biomass represents 74% of the energy produced, thus making it the most important "renewable" source of energy (for the problem of counting bioenergy as renewable see Harjanne and Korhonen 2019). Despite the fact that bioenergy is often viewed as a renewable, it causes significant greenhouse gas emissions (see Searchinger et al. 2018; Letter from Scientists 2018; Vadén et al. 2019).

7.3 A Twofold Threat to the Regime: Carbon Sinks and EU Regulation

The vision of Finnish bioeconomy as sustainable started to crack when the bioeconomy regime was challenged by climate science. The Finnish bioeconomy strategy (2014) and the bioeconomy plans associated with it involved increasing forest harvesting. In addition, the centre-right government's Energy and Climate Strategy (Huttunen 2017) was also based on increasing the annual harvesting of forests—from 65 million to a record-breaking 80 million cubic metres. As mentioned above, increasing harvesting—the material basis of the forest bioeconomy—faced very little criticism when the Finnish bioeconomy strategy was first introduced. Thus, the cornerstones of Finland's "actually existing forest bioeconomy" (the chemical forest industry and bioenergy) were generally accepted as environmentally sustainable solutions.

For quite some time, therefore, the use of forest biomass had been considered sustainable and was even promoted for climate reasons: harvested forests were to be replaced by new forest growth, which would soak up the carbon emissions associated with harvesting. This argument promotes forest-based bioenergy, for example as an attractive and renewable replacement for fossil fuel. However, the situation looks different when the rapid time span of climate mitigation is taken into consideration. Harvesting decreases the size of forest carbon sinks, and when wood is used for short-term products, such as paper, pulp or bioenergy, carbon is released immediately or relatively quickly into the atmosphere. New boreal forests take decades, in some cases even more than a century, to store the carbon released by harvesting (e.g. Sievänen et al. 2014; Soimakallio et al. 2016; Public Statement 2017). Importantly, it takes more than two decades for newly planted forests to even start beginning to store significant amounts of carbon. If there is pressure to increase harvesting, this makes short-term carbon neutrality targets even harder to achieve. These facts were brought to the public's attention from 2014 onwards by climate researchers, the FCCP and, in March 2017, by a public statement signed by 68 Finnish researchers (see Public Statement 2017). Together, this evidence questioned the sustainability of the bioeconomy regime.

However, the issue of carbon sinks was not the only black cloud that was gathering above the Finnish bioeconomic imaginary. The EU was also reconsidering the principles behind its climate policy, and the regulation of how member states use their lands and forests. The EU "land use, land use change and forestry" regulation (LULUCF) draws up binding commitments for each member state to regulate its emissions from land use. It is the latest set of regulations in the EU climate and energy policy framework, and LULUCF policies are to be enshrined in EU law by 2021.

This upgrading of LULUCF monitoring made land use emissions an important political issue, especially for member countries with large forestry sectors. In order to monitor member states' land use, a "forest reference level" (an estimate of annual average net emissions from managed forest land) was set. Eventually, the EU chose the period between 2000 and 2009 as the reference level. In Finland, the reference level was viewed as unfavourable because the decline of the chemical forest industry had led to reduced harvesting during the 2000s. The planned increase in harvesting from 2015 onwards would significantly decrease the size of carbon sinks and increase net emissions from the land use sector. Finland pointed out that the proposed reference level treated the Nordic member states unequally: following historical harvesting levels placed Sweden into a more favourable position. During the whole LULUCF negotiation process, the official line of the centre-right government was that member states, not the EU, should have the final say on how forest biomass is used.

Consequently, and rather unpredictably, the Finnish bioeconomy regime faced challenges on two fronts. On the one hand, this led forestry actors to engage in public discussions about carbon sinks. On the other, it led the interests of Finnish forestry to be defended at the EU level. In the next three sections, I analyse the key public discursive strategies that different forestry actors used to frame the discussion about the role of forests in climate mitigation in order to support the existing bioeconomy regime. By forestry actors, I mean organisations with close ties to the interests of Finnish forestry, mainly the representatives of the Finnish Forest Industries, the Central Union of Agricultural Producers and Forest Owners and the Central Party. I distinguish between three chronological periods in the public debate: the regime under shock, the battle in the EU and stabilisation of the regime.

7.4 The Regime Under Shock

A public statement, undersigned by 68 Finnish researchers, released in March 2017, caused a public "storm" (Hukkinen et al. 2017) that shocked the bioeconomy regime. There was nothing fundamentally new in the substance of the statement: it merely stated that increasing harvesting caused a threat both to carbon sinks and to forest biodiversity—both facts had already been acknowledged in several scientific publications. This time, however, the message was brought to the public not by individual researchers or by the FCCP, which already had established role in Finnish climate science communications, but collectively by researchers from a broad spectrum of environmental studies.

The shockwaves sent through the regime led advocates of the bioeconomy regime to put forward a set of aggressive arguments.[1] However, the researchers' statement provoked these sentiments even before it had been published. At a time when the statement was still circulating among researchers with the aim of gaining signatories, it was provided to the media and was dismissed by forestry actors. For example, Katri Kulmuni, the then vice-chair, and later leader of the Centre Party, referred to the authors and potential signatories of the statement as "unpatriotic". When the statement was published, a leader of the Central Union of Agricultural Producers and Forest Owners (MTK) dismissed it as "a political pamphlet" that "ignores scientific facts" and focuses on a "narrow perspective and a short time period". One forestry leader wanted to teach researchers a lesson about the growth of forests by sending them to "a course where they can be taught how carrots grow". Another leader chose even more innovative phrasing by calling the statement "forest Trumpism" and recommending that the "researchers, who published the clearly political pamphlet, should stop prattling and grab a chainsaw instead". *Maaseudun Tulevaisuus*, a newspaper closely associated with the forestry sector, questioned where the researchers who had facilitated and signed the statement received their funding.

Although the statement made carbon sinks an unavoidable issue in Finnish climate politics, it was also used to create the impression that two camps of researchers existed with different opinions on the role of forests in climate mitigation. In practice, this was not the case. Researchers

---

[1] The statements analysed here are drawn from forthcoming wider analysis of the Finnish carbon sink debate. The arguments appeared in three major Finnish mass mediums, *Yleisradio* (Finnish National Broadcasting Company), *Helsingin Sanomat* (the biggest daily newspaper) and *Maaseudun Tulevaisuus* (a newspaper closely associated with the Centre Party and the forest industry). Katri Kulmuni's statement is from her Facebook post (16 March 2017).

are clearly unanimous about the key fact that increasing harvesting will reduce carbon sinks during the period in which climate mitigation can still have an impact. However, political differences exist when people do not accept climate politics or the rapid timetable for mitigation as a steering framework for political decisions or scientific inquiry. Nevertheless, forestry actors focused their attention on the legitimacy of the statement despite the fact that the same arguments about carbon sinks and forest use had already been expressed in several research publications, including those by the Finnish Climate Change Panel (FCCP 2015) and the European Academies' Science Advisory Council (Aszalós et al. 2017), as well as a public letter signed by 800 European scientists (Letter from Scientists 2018).

During this period, the bioeconomy regime set aside its internal differences and organised a front against a common threat. In the beginning of the period characterised by the centre-right government, things had been different: forest actors debated openly about how and who should be able to use scarce forest resources. In 2015, the forest industry claimed that the higher value-adding chemical forest industry would face resource shortages if the share of forest-based bioenergy were to be increased. In contrast, the forest owners (MTK) provided reassurances that there was enough wood in Finnish forests to implement all of the planned bioeconomic strategies. Simultaneously, the credibility of the centre-right government's bioeconomy target, 100,000 new bioeconomy jobs, was openly questioned by forestry actors. These opinions were an expression of an internal struggle over the bioeconomic monetary flows promised by the new government.

7.5 The Battle in the EU

In June 2016, the EU Commission presented the LULUCF legislative proposal. In principle, it followed contemporary scientific findings about the role of forests in climate mitigation, and, as such, it was directed at instantly blocking forest policies that led to increased harvesting and reducing the size of carbon sinks. In Finland, the forest bioeconomy regime viewed the proposal as a declaration of war. When the forest regime realised that Finnish lobbying in the EU had failed terribly, it began to search for the "guilty parties". Fears were expressed that the proposal would clash with the country's bioeconomy plans and forestall investments in new pulp factories.

A new front was built. One member of the European Parliament, who had been engaged deeply in the LULUCF process, declared that "a spirit of Winter War is needed" to amend the Commission's proposal. The bioeconomy regime started a campaign which, in the history of Finnish EU lobbying, can be viewed as "exceptionally voluminous". Most Finnish members of the EU Parliament, the centre-right government and different actors from the forestry sector used all of their power to influence the situation in the name of the "interests of the Fatherland" (as it was often portrayed in the media). In addition, political alliances were sought from other member states; Sweden, with its large forest sector, being the most important. One year later, in July 2017, despite all of the war rhetoric, the European Parliament Committee on the Environment (ENVI) voted for even tighter regulation of the LULUCF sector. ENVI entailed annual surveys of the trajectory of member states' carbon sinks and restricting unfavourable forest use immediately, not at some point in a possible future—again, a position that was in line with scientific consensus on the role of forests in climate mitigation.

This was followed by a final (and successful) round of lobbying by Finland. In September 2017, the European Parliament finally approved the LULUCF legislation; however, Finland, supported by other countries with large forestry sectors, such as Sweden, had lobbied successfully for a crucial change to the LULUCF proposal. The original proposal's focus on immediate change to forest use and carbon sinks was altered in favour of a concentration on the long-term perspective, which left the door open to immediately increasing harvesting in certain member states if carbon sinks were expanding in the EU as a whole. The bioeconomy regime celebrated the vote as a historical victory. The CEO of Finnish Forest Industry, interviewed by *Yleisradio* (the Finnish Broad Casting Company, 13 September 2017), expressed gratitude for the successfully conducted "national endeavour" and described the political importance of the LULUCF decision for the world in honest terms: "The world is not getting better here. But we are now blocking decisions that would be totally unreasonable for Finland and bad for our economic development and forest industry".

7.6 Stabilising the Regime

After the battle had been won at the EU level, a more nuanced discussion on behalf of the forestry regime followed. Forest actors strived to stabilise the faltering bioeconomy regime. Interestingly, during this period, the concept of bioeconomy appears less in public discussion. Further research is needed to analyse why this happened, but the focus of the forest debate obviously shifted from abstract promotion of the bioeconomic imaginary to more concrete and conflictual issues that eventually constituted the future of the bioeconomy regime. In the wide spectrum of public debate that took place between 2017 and 2019, three dominant discursive strategies continued to confront the scientific arguments on the role of forests in climate mitigation. Forestry actors focused on using the frame of sustainable development, emphasised the special characteristics of Finland as *the* land of the forest industry and drew attention to the growth of forests.

First, in answer to the scientists' point about ecological sustainability, forest actors emphasised that it was only one aspect of a broader sustainability perspective and that it was essential that economic and social factors were provided with an equal level of recognition in the future of Finnish forestry. Thus, the forestry actors presented the famous three pillars of sustainability (ecological, social and economic development) as equally important. In contrast, the researchers concerned about the diminishing size of carbon sinks argued that the short period available for climate mitigation meant that planetary limits needed to be prioritised and that social and economic development would have to be adjusted accordingly (see Hukkinen et al. 2017).

A second strategy is familiar from any discussion about climate change: the smallness of a country is held as a justification for a moderate level of climate action, which, after all, is a global problem that no country can solve alone. Thus, the argument goes, the question about

Finnish forest carbon sinks is irrelevant in the bigger picture when countries such as China and India are largely responsible for climate change. Furthermore, forest actors stressed that Finnish forest bioeconomy was the most sustainable in the world and that if harvesting were to be restricted, the production of bioeconomy products (i.e. pulp and paper) would be relocated to countries with less stringent environmental regulation. Thus, the production of bioeconomy products in Finland was framed as an act of climate mitigation.

Third, and most importantly, forestry actors focused the carbon sink discussion on the expansion of forests. As stated above, a favourable decision on the LULUCF regulation was extremely important for Finland. In its final form, LULUCF does not penalise countries in which increased harvesting is causing an annual decrease in the size of their carbon sinks *if* carbon sinks are increasing annually throughout the EU as a whole and in all member countries in the long term.

During the period in which the LULUCF legislation was being drawn up, the Finnish Natural Resource Institute remodelled the growth of Finland's forests. At the end of 2018, the Institute published new results demonstrating that forests in Finland were expanding significantly. This pushed up the previous estimate of "economically sustainable" harvesting (and which enabled harvesting to be increased) if the immediate threat of a loss of carbon sinks caused no ramifications. This has been the case since the implementation of the LULUCF regulation. In addition, the estimate indicated that forest carbon sinks had expanded dramatically. Between 2015 and 2024, for example, forest carbon sinks were now said to have grown from 16.5 megatons (Mt) CO_2 to close to 40 Mt CO_2; these figures put the carbon sinks at more than twice the size estimated two years earlier. The unexpected increase in growth was said to be a result of better forest management, healthier saplings and the effects of climate change.

These results were interpreted by the bioeconomy regime primarily as evidence of good forest management. Thus, now that forests had grown much more than expected, there was room for increased harvesting—in other words, the regime could continue with the expansion frame without complying with the transformation frame. The estimates produced by the Forest Institute and its interpretation of the bioeconomy regime did not satisfy climate scientists. In February 2019, the Finnish Climate Change Panel published a report (FCCP 2019) that provided five different models of both the growth of Finland's forests and the development of carbon sinks in the country. The report found that only one model supported the idea that forests could be simultaneously harvested and preserved as carbon sinks. This model was known as the MELA model and was the one that the Forest Institute (LUKE) was using. Furthermore, the Climate Panel report stated that all of the other models demonstrated that increased harvesting would lead carbon sinks to decrease in size and that this would continue to be the case for decades to come. For this reason, the Panel recommended that "the cleverest thing would be to decrease the level of harvesting". This case demonstrates just how complicated, technical and, perhaps, political, estimating climate impact can be.

7.7 Conclusion

When public awareness of the multifarious environmental crisis increases, and climate regulation tightens, it seems obvious that the sustainability of the bioeconomy imaginary will be critically evaluated. This occurred in Finland rather suddenly when climate scientist questioned the path set by the regime and, at the same time, the EU aimed to draw up a more ambitious form of climate regulation. The reaction by Finnish forestry was forceful. The bioeconomy regime engaged aggressively with public debate and harnessed all of its power to influence EU climate regulation in the interests of the Finnish forestry sector. Towards the end of the period under analysis, the strategies of forestry actors became more nuanced and focused on the growth of Finland's forests.

In global climate politics, "a small and medium-sized power" like Finland, can gain a bigger role than its position in world politics might imply. The capability of Finland to effectively lobby at the EU level for forest policies that served the interests of its forestry sector complicates the achievement of ambitious climate targets in Finland and in the rest of Europe. Importantly, it is impossible to rule out that this will have a major impact on the climate: if Finland is capable of lobbying for international policies that enable increased forest harvesting in a period in which rapid climate mitigation is essential, what would prevent other countries from following suit?

I have analysed the evolution of bioeconomy in Finland as a novel forest policy regime. The bioeconomic imaginary relocated the traditional forestry sector with its strong green image to the core of Finnish society. It is too early to assess the resilience and future of this relatively new forest regime. However, the case of Finnish bioeconomy demonstrates how the success of the bioeconomic imaginary in national terms requires favourable political conditions; this makes the bioeconomy regime dependent on existing political trends. The approach of the new Finnish left-green government, which was formed in May 2019, to forest bioeconomy and climate politics is substantially different to the policies adopted by the centre-right government in 2014. At this time, the bioeconomy was booming and nobody, with the exception of specialists from this field, had even heard of "carbon sinks". If the bioeconomy imaginary loses national ground when the political winds turn, the attractiveness of the transnational bioeconomy might also weaken relatively quickly. For this reason, a more sophisticated analysis of the political economy of bioeconomy is needed: critical research should evaluate which tendencies of the bioeconomy project, if any, are actually sustainable at the international level in the face of changing political conditions.

References

Ahlqvist, T., & Sirviö, H. (2019). Contradictions of Spatial Governance: Bioeconomy and the Management of State Space in Finland. *Antipode, 51*(2), 395–418.

Aszalós, R., Ceulemans, R.J., Glatzel, G., Hanewinkel, M., Kakaras, E., Kotiaho, J., et al. (2017). *Multi-functionality and Sustainability in the European Union's Forests, EASAC Policy Report* (No. 32). Halle (Saale): European Academies Science Advisory Council (EASAC). https://easac.eu/fileadmin/PDF_s/reports_statements/Forests/EASAC_Forests_web_complete.pdf. Accessed 30 Sept 2020.

Birch, K. (2006). The Neoliberal Underpinnings of the Bioeconomy: The Ideological Discourses and Practices of Economic Competitiveness. *Genomics, Society and Policy*, 2(3).

Birch, K., & Tyfield, D. (2013). Theorizing the Bioeconomy: Biovalue, Biocapital, Bioeconomics or…What? *Science, Technology, & Human Values*, 38(3), 299–327.

Bugge, M., Hansen, T., & Klitkou, A. (2016). What Is the Bioeconomy? A Review of the Literature. *Sustainability*, 8(7), 691.

Donner-Amnell, J., Lehtinen, A., & Saether, B. (2004). Comparing the Forest Regimes in the Conifer North. In A. Lehtinen, J. Donner-Amnell & B. Sateher (Eds.), *Politics of Forests: Northern Forest-industrial Regimes in the Age of Globalization* (pp. 255–284). Ashgate: Alfershot.

Eloranta, J., Golson, E., Hedberg, P., & Moreira, M. C. (Eds.) (2018). *Small and Medium Powers in Global History: Trade, Conflicts, and Neutrality from the 18th to the 20th Centuries*. New York: Routledge.

Entman, R.M. (2007). Framing Bias: Media in the Distribution of Power. *Journal of communication*, 57(1), 163–173.

FANC (2014). Ollako vaiko eikö olla: Askelkuvio biotaloudelle. https://www.sll.fi/app/uploads/2018/10/biotalous_raportti_sll_2014.pdf. Accessed 15 Nov 2019.

FCCP (2015). Metsien hyödyntämisen ilmastovaikutukset ja hiilinielujen kehittyminen. https://www.ilmastopaneeli.fi/wp-content/uploads/2018/10/Metsien-hyodyntamisen-ilmastovaikutukset-ja-hiilinielujen-kehittyminen.pdf. Accessed 15 Nov 2019.

FCCP (2018). Ilmastopaneelin näkemykset pitkän aikavälin päästövähennystavoitteen asettamisessa huomioon otettavista seikoista. https://www.ilmastopaneeli.fi/wp-content/uploads/2018/10/Ilmastopaneelin-muistio_hyvaksytty_4.6.2018.pdf. Accessed 15 Nov 2019.

FCCP (2019). Skenaarioanalyysi metsien kehitystä kuvaavien mallien ennusteiden yhtäläisyyksistä ja eroista. https://www.ilmastopaneeli.fi/wp-content/uploads/2019/02/Ilmastopaneeli_mets%C3%A4mallit_raportti_180219.pdf. Accessed 15 Nov 2019.

Finnish Bioeconomy Strategy (2014). Kestävää kasvua biotaloudesta – Suomen biotalousstrategia. https://biotalous.fi/wp-content/uploads/2014/07/Julkaisu_Biotalous-web_080514.pdf. Accessed 15 Nov 2019.

Goven, J., & Pavone, V. (2015). The Bioeconomy as Political Project: A Polanyian Analysis. *Science, Technology, & Human Values*, 40(3), 302–337.

Harjanne, A., & Korhonen, J.M. (2019). Abandoning the Concept of Renewable Energy. *Energy Policy*, 127, 330–340.

Harjuniemi, T. (2019). Reason over Politics: The Economist's Historical Framing of Austerity. *Journalism Studies*, *20*(6), 804–822.

Hukkinen, J.I., Kotiaho, J.S., & Vesala, T. (2017). Kirje, joka nostatti myrskyn [A Letter That Caused a Storm]. *Alue ja Ympäristö*, *46*(1), 46–51.

Huttunen, R. (Ed.) (2017). *Valtioneuvoston selonteko kansallisesta energia-ja ilmastostrategiasta vuoteen 2030*. Helsinki: Publications of the Ministry of Economic Affairs and Employment 4/2017. http://julkaisut.valtioneuvosto.fi/bitstream/handle/10024/79189/TEMjul_4_2017_verkkojulkaisu.pdf?sequence=1&isAllowed=y. Accessed 15 Nov 2019.

Kotilainen, J., & Rytteri, T. (2011). Transformation of Forest Policy Regimes in Finland Since the 19th Century. *Journal of Historical Geography*, *37*(4), 429–439.

Kröger, M., & Raitio, K. (2017). Finnish Forest Policy in the Era of Bioeconomy: A Pathway to Sustainability? *Forest policy and Economics*, *77*, 6–15.

Letter from Scientists (2018). Letter from Scientists to The EU Parliament Regarding Forest Biomass. http://www.pfpi.net/wp-content/uploads/2018/04/UPDATE-800-signatures_Scientist-Letter-on-EU-Forest-Biomass.pdf. Accessed 15 Nov 2019.

Meyer, R. (2017). Bioeconomy Strategies: Contexts, Visions, Guiding Implementation Principles and Resulting Debates. *Sustainability*, *9*(6), 1031.

Peltomaa, J. (2018). Drumming the Barrels of Hope? Bioeconomy Narratives in the Media. *Sustainability*, *10*(11), 4278.

Public Statement (2017). Public Statement of 68 Researchers. https://bios.fi/publicstatement/publicstatement240317.pdf. Accessed 15 Nov 2019.

Rockström, J., Gaffney, O., Rogelj, J., Meinshausen, M., Nakicenovic, N., & Schellnhuber, H.J. (2017). A Roadmap for Rapid Decarbonization. *Science*, *355*(6331), 1269–1271.

Searchinger, T.D., Beringer, T., Holtsmark, B., Kammen, D.M., Lambin, E.F., Lucht, W., & van Ypersele, J.P. (2018). Europe's Renewable Energy Directive Poised to Harm Global Forests. *Nature communications*, *9*(1), 3741.

Sievänen, R., Salminen, O., Lehtonen, A., Ojanen, P., Liski, J., Ruosteenoja, K., et al. (2014). Carbon Stock Changes of Forest Land in Finland Under Different Levels of Wood Use and Climate Change. *Annals of forest science*, *71*(2), 255–265.

Siiskonen, H. (2007). The Conflict Between Traditional and Scientific Forest Management in 20th Century Finland. *Forest Ecology and Management*, *249*(1–2), 125–133.

Soimakallio, S., Saikku, L., Valsta, L., & Pingoud, K. (2016). Climate Change Mitigation Challenge for Wood Utilization. The Case of Finland. *Environmental Science & Technology, 50*(10), 5127–5134.

Vadén, T., Majava, A., Toivanen, T., Järvensivu, P., Hakala, E., & Eronen, J.T. (2019). To Continue to Burn Something? Technological, Economic and Political Path Dependencies in District Heating in Helsinki, Finland. *Energy Research & Social Science, 58*, 101270.

Open Access This chapter is licensed under the terms of the Creative Commons Attribution 4.0 International License (http://creativecommons.org/licenses/by/4.0/), which permits use, sharing, adaptation, distribution and reproduction in any medium or format, as long as you give appropriate credit to the original author(s) and the source, provide a link to the Creative Commons license and indicate if changes were made.

The images or other third party material in this chapter are included in the chapter's Creative Commons license, unless indicated otherwise in a credit line to the material. If material is not included in the chapter's Creative Commons license and your intended use is not permitted by statutory regulation or exceeds the permitted use, you will need to obtain permission directly from the copyright holder.

8

Sugar-Cane Bioelectricity in Brazil: Reinforcing the Meta-Discourses of Bioeconomy and Energy Transition

Selena Herrera and John Wilkinson

8.1 Introduction

In Brazil, more than 80% of the electrical matrix is composed of renewable sources. Sugar-cane bioelectricity from residues occupies third place behind hydroelectric and natural gas power plants if its use to power sugar factories is taken into consideration. However, it comes in fourth place, slightly behind wind, in terms of supplies to the national grid (EPE[1] 2019). The expected decrease in the supply of hydroelectric energy, which accounts for over 60% of the energy mix (MME[2] 2019a), and the rapid increase in the supply of natural gas for power generation (ANP[3] 2018) mean that additional power generation from renewable

---

[1] Empresa de Pesquisa Energética: Energy Research Office
[2] Ministério de Minas e Energia: Ministry of Mines and Energy.
[3] Agência Nacional do Petróleo Gás Natural e Biocombustíveis: National Agency of Petroleum, Natural Gas and Biofuels.

S. Herrera (✉) · J. Wilkinson
Federal Rural University of Rio de Janeiro, Rio de Janeiro, Brazil

© The Author(s) 2021
M. Backhouse et al. (eds.), *Bioeconomy and Global Inequalities*,
https://doi.org/10.1007/978-3-030-68944-5_8

sources is a critical factor in the clean energy transition in Brazil. Global consensus around the need for an energy transition can provide powerful stimuli for institutional changes in favour of bioenergy promotion. This chapter analyses how meta-discourses regarding energy transitions and the bioeconomy influence the expansion of sugar-cane bioelectricity in Brazil.

Brazil has already become a key reference in global energy geopolitics through its sugar-cane ethanol fuel programme (Wilkinson and Herrera 2010). Bioelectricity, produced from bagasse and straw resulting from the harvesting and treatment of sugar-cane, has now become the third most important by-product of the sugar-cane industry, with 46% of producers using it to power their factories and 54% supplying energy to the national grid (UNICA[4] 2019). The emergence of actors from new energy combinations, including natural gas and new renewable energies (namely wind and solar), is having an impact on the governance of sugar-cane bioelectricity. As both a substitute for petrol and a food commodity, sugar-cane production in Brazil has to negotiate a complex institutional and market environment covering fuel, food and agriculture (Kuzemko et al. 2016). Bioelectricity now adds a new component to this complexity as it involves a competitive market environment that includes both fossil fuel (coal, diesel and natural gas) and alternative renewables (biomass, hydraulic, wind and solar).

The energy transition is generally understood as a fundamental structural change in the energy sector of a particular country that involves the phasing out of fossil energy sources (Hauff et al. 2014) and their replacement with renewables. The bioeconomy is based on the idea of applying biological principles and processes to all sectors of the economy together with the increasing replacement of fossil-based raw materials in the economy with bio-based resources and principles (Dietz et al. 2018). Moreover, it also involves the promotion of bioenergy as a renewable energy source. To replace fossil energies in the electric sector, sugar-cane bioelectricity must not only become competitive in terms of costs, but must also face up to competition from other renewable sources, especially wind and solar, in the formulation of policy-promoting instruments.

---

[4]União da Indústria de Cana-de-Açúar: Brazilian Sugarcane Industry Association.

The following sections present the analytical framework of the arguments put forward in this chapter, which rely on the notion of socio-technical transitions as viewed from a multi-level (macro/landscape, meso/regime and micro/niche) perspective. Sugar-cane bioelectricity is then situated within the macro-context of the energy transition and the regulatory and market forces governing the sugar-cane sector. The chapter then explores the potential of sugar-cane bioelectricity to establish itself as a dynamic niche within the overall energy transition.

8.2 The Analytical Framework

In the "multilevel perspective on transitions", transitions are considered nonlinear processes that result from the interaction between three levels of development: the landscape (macro-level), the regime (meso-level) and niches (micro-level) (Geels and Schot 2007). The notion of the socio-technical regime refers to the predominant set of routines or practices that actors and institutions adopt and that create and reinforce a given technological system (Rip and Kemp 1998). Although specific definitions of "regime" may differ, an essential characterization refers to its dominant position and its reproduction of dominant structures in a particular social system. As such, a regime is, by definition, associated with "power", "dominance" and "vested interests" (Avelino and Wittmayer 2016). The landscape (the convergence around meta-discourses), in contrast, is the macro-level that can influence the dynamics of both the regime and niche levels (Rip and Kemp 1998) and that actors cannot influence in the short term (Grin et al. 2010).

In the multilevel perspective, changes in the socio-technical regime develop out of selection pressures arising from the landscape and technological niches, as well as from within the regime itself (Geels and Schot 2007; Smith et al. 2005). Among the pressures from the landscape on a global scale, meta-discourses correspond to the emergence of new beliefs or political challenges. Regime transition is promoted when the selection pressures resulting from meta-discourses reinforce each other and when resources such as investments, capacities and knowledge are coordinated with these selection pressures (Hall 2010).

Actors with different powers and interests within the regime play critical roles in shaping the discourse, setting the policy agenda and framing, supporting or suppressing niches, or simply lobbying to obstruct or promote legislation (Andrews-Speed 2016). Thus, the established energy system and the associated institutional structure often prevent the adoption of potentially superior alternatives (Foxon and Pearson 2007). From this perspective, meta-discourses can be harnessed by sugar-cane bioelectricity producers to exert pressure on the established regime in order to bring about institutional changes that are favourable to the expansion of bioelectricity. The extent to which this is the case in Brazil is the focus of this chapter.

8.3 The Landscape: The Meta-Discourses of Bioeconomy and Energy Transition

The "landscape" represents the broader political, social and cultural values and institutions that form the deep structural relationships of a society and change slowly (Foxon et al. 2010). Pülzl et al. (2014) view discourses—a set of concepts that become transformed into a particular set of practices (Hajer 1995)—as generally very stable and rarely changing overnight. The concepts of energy transition and bioeconomy are treated in this chapter as two global meta-discourses that have recently emerged from previous debates and that will have an impact on global policies in the coming decades.

Beginning as far back as the 1970s, a global convention has become consolidated around the need to move towards an energy mix based on renewable energies. This is called the clean energy transition by the European Union (EU 2019). This convention gained force with the United Nations Conference on Environment and Development held in Rio de Janeiro in 1992, as well as with the 2015 Millennium Development Goals (MDGs) and the 2030 Sustainable Development Goals (SDGs). The discourses of energy transition and bioeconomy are thus rooted in the notion of sustainability and its three components (economic, social and environmental) that are aimed at facing up to global energy challenges (Dubash and Florini 2011; Goldthau 2013).

In Brazil, meta-discourses on climate change and sustainable development have been translated into policies such as the National Policy on Climate Change (implemented in 2009) with commitments until 2020 (Law 12,187/2009). In 2016, Brazil ratified the SDGs and the Paris Agreement, linking its commitments to energy and the bioeconomy. In addition to reducing greenhouse gas emissions, Brazil committed to increasing the share of renewable energies (wind, biomass and solar in addition to hydropower) in its supply of electricity to at least 23% by 2030, and the share of sustainable bioenergy in the energy mix to 18% (MRE[5] 2015). Consistent with these environmental commitments, but also for reasons of energy security and industrial development, the National Biofuels Policy (RenovaBio—Law 13,576) was launched in 2017 to expand Brazil's bioenergy production capacity and was supported by an environmental certification system, the Biofuel Decarbonization Credit or CBIO. This institutional change provided new impetus to the sugar-cane industry.

In the case of Brazilian biofuels, a historical perspective shows that the influence of international discourses and the prospects of demand were decisive and that these translated into new policies based on the promotion of sustainable practices (e.g. zoning law and mechanization of the harvest) and bio-diplomacy (Wilkinson 2014). In terms of the market, the meta-discourses of energy transition and bioeconomy influenced the adoption of sustainability certifications by the biofuels sector to embrace the possibility of exports, as in the case of the European voluntary schemes. By using the approaches proposed by van Dam et al. (2008) for biomass sustainability certifications, Herrera (2014) argues that, while the private sector and civil society have the power to act directly on the private certifications criteria, states participate through governmental networks (Slaughter 2005) on the definition of national regulations and global principles.

However, signs of discontinuity in climate policy, the Brazilian government's weakening of policies implemented in 2019 to prevent and combat deforestation and the expansion of electricity generation from non-renewable sources threaten to divert the country's path to fulfilling

---

[5]Ministério das Relações Exteriores: Ministry of Foreign Affairs (Itamaraty).

its climate change mitigation commitments (CMA[6] 2019). The efficacy of established meta-discourses is currently being challenged by the Bolsonaro Government in Brazil, and this is part of a broader tendency that is challenging notions of sustainability and that will illuminate the leeway that national policies have to flout the global practices that arise from conventions based on these meta-discourses.

Until the emergence of global meta-discourses on sustainability and climate change, bioenergy governance in Brazil was characterized by national policy and the strength of the sugar-cane agro-industry lobby. The consolidation of these meta-discourses has led Brazilian bioenergy governance to become open to decisions taken at the international level and to new actors interested in the market opportunities that arise from them. As Bernstein and Cashore (2012) argue, it is not only about the "compliance" or the "effectiveness" of the meta-discourses in Brazil, but also about their "influence", since the latter includes "the combined effects of the international and transnational efforts on domestic or firm policies and practices". Both the sugar and the oil and gas markets in Brazil are already part of a global governance. Although there is only an incipient market for ethanol in the USA and Europe despite its lower emission of greenhouse gases (Herrera 2014; NovaCana 2020), the influence of global governance is evident in the sustainability claims put forward to defend in Brazil's sugar-cane ethanol (Wilkinson 2014).

Through the lens of the energy transition and bioeconomy meta-discourses, sugar-cane bioelectricity implies, on the one hand, the substitution of fossil energy sources. However, it also involves other renewable and sustainable energy actors that have complementary interests in the promotion of alternative sources, but which can also be competitors in the promotion and design of political instruments, especially wind and solar, both of which are undergoing significant growth in Brazil. Faced with the rigidities of established interests and pathways associated with fossil energies and ethanol, the meta-discourses have also stimulated other technological options to compose the new energy mix, which poses a direct challenge to sugar-cane bioelectricity.

---

[6]Comissão de Meio Ambiente: Brazilian Environment Commission.

8.4 An Emerging Renewable Electricity Regime

Brazil's energy sector presents a complex architecture of actors that splits responsibilities for energy regulation into the electrical power and the oil and gas sectors, all of which are subordinated to the Ministry of Mines and Energy but are separated from the environmental sector. Streeck and Thelen (2005) and Nunes (1997) argue that the development of Brazil's state bureaucracies has been extremely unequal and heterogeneous. The institutional changes adopted along the national development path (Colomer and Queiroz 2019) have resulted in an energy sector that is not regulated holistically and that is strongly affected by other national political issues (Hauff et al. 2014). During its previous transitions, Brazil has focused its priority on the national security of energy supply and demand and on economic development. From a historical perspective, bioethanol expansion is congruent with an energy policy based on self-sufficiency and local industrialization (Rodríguez-Morales 2018). Moreover, the Brazilian state has played the role of market maker, regulator and motivator for the ethanol market and for biodiesel (Herrera 2014).

With the adoption of ethanol as fuel, the interaction of the energy and food markets accelerated the institutionalization of the sugar-cane industry as part of the socio-technical system of road transport (Rodríguez-Morales 2018). By developing the bioelectricity market, the sugar-cane industry is also institutionalized as part of the socio-technical system of the electricity sector. The Brazilian electricity sector has historically been structured around the planning of the country's hydroelectric resources. As of the 1970s, the state fostered rapid industrialization and economic development through hydroelectric energy (de Oliveira 2007). In the 1990s, privatization and restructuring transformed the energy sector and state-owned companies, introducing competition and attracting more foreign investors (Bradshaw and de Martino Jannuzzi 2019). Since 2003, the electricity sector has favoured the commercialization of wind and solar energy, and has done so as part of a strategic plan based on distributed electricity generation, the implementation of smart grids and the diversification of energy sources (Garcez 2017). In the past, fossil fuel-based thermoelectric plants were built to compensate for the lack of hydroelectricity during periods of low rainfall. Whereas hydraulic energy faced progressively greater costs and environmental constraints, e.g. in the fragile ecosystem of the Amazon, natural gas availability increased by 85% between 2010 and 2017 (ANP 2018).

On the one hand, the meta-discourse of sustainability, climate and the energy transition stimulates wind and solar energy because they are alternatives to fossil fuel, but they are also supported by interests linked to natural gas. The generation of energy using natural gas has attracted more attention as it has lower specific CO_2 emissions and greater operational flexibility than energy produced by coal and oil (Khallaghi et al. 2020). At the same time, it is also independent of climate variations, which brings reliability gains to the system (TCU[7] 2018). The switch from coal and oil to natural gas could contribute to a cleaner energy production. However, gas is not a clean source of energy since it results from oil exploration, a global energy system, and Brazil has created expectations of exporting gas on global markets (IEA—International Energy Agency 2019). However, it could also improve the viability of intermittent renewable energy sources, such as wind and solar energy sources, (ANP 2018) until an energy storage technology were to be made available at a competitive price, and, therefore, provide for national energy security. Renewable energy sources and natural gas are predicted to account for 85% of energy growth, representing 15 and 26% of the primary energy consumption in 2040, respectively—oil being in first place with 27% (BP 2019). This transition, however, responds to the market appeal of low-cost supplies of both and an increasing global availability of gas, aided by the growing supplies of liquefied natural gas (ibid.). The discovery of offshore fields (*Pré-Sal*) in 2006 confirmed Brazil as having one of the largest oil and gas reserves in the world (Goldemberg et al. 2014). The "new gas market" (MME 2019b), implemented in June 2019 in the context of a liberalization of energy markets for foreign companies, represents a key component in the new energy transition in Brazil, combined with wind and solar sources.

---

[7]Tribunal de Contas da União: Federal Audit Court.

Although there are other potential by-products, bioelectricity has emerged as the third major co-product from sugar and ethanol milling, and it is produced from the bagasse and straw that results from the harvesting and treatment of sugar-cane. Each tonne of sugar-cane processed in the mill produces around 270 kg of bagasse and 155 kg of straw—these figures are expected to increase with the introduction of the new variety of energy-cane (EPE 2018), in addition to 72 kWh of bioelectricity (EPE 2017). Bioelectricity generation and its efficiency, however, depend on technological choices (de Souza and de Azevedo 2006). The history of sugar-cane has already shown the importance of government financial support from Brazil's National Development Bank (BNDES) (Wilkinson 2015). Political support alone does not necessarily provide sugar-cane with greater access to energy markets, as this also depends on the government's interest in expanding bioenergy and its application of regulations and incentives for technological improvement.

The commercialization of sugar-cane bioelectricity has been stimulated ab initio through public policies based on energy security concerns, as is the case of the market for ethanol. The "Programa Nacional do Álcool – *Proálcool*" or National Alcohol Plan was created in 1975 and characterized by a high level of market intervention. Suddenly, bagasse was no longer a nuisance that had to be burned but a marketable good that was progressively adopted by the industries close to the distilleries. A prolonged drought in 2001 resulted in a national energy crisis and this provided the first major political impetus for the development of non-hydroelectric renewable energy sources. In this context, the first institutional change in favour of bioelectricity came in 2002 in the form of Proinfa—the Programme to Stimulate Alternative Sources. Its objective was to increase the proportion of alternative renewable sources used to produce electricity (small hydroelectric plants, wind power plants and biomass thermoelectric projects—solar energy was not included at this time) compared to the large hydraulic reservoirs. In 2003, the national wind industry began to receive financing from the BNDES and since then it has become competitive and focused on an industrial policy that advocates the local components production. Between 2003 and 2009, ethanol production increased annually by some 13%, thanks to national and global investments that poured into the sector in addition to the huge support provided by the BNDES. The boom in resources contracted from the BNDES for cogeneration in 2018 highlights the importance of the RenovaBio policy for bioelectricity. Meanwhile, the first BNDES financing of solar energy was only approved in 2017 and the solar sector still depends on imports of materials (Costa 2020).

In addition to direct incentives (mainly from the BNDES and tax exemptions for wind energy), the main public policy instrument to regulate the energy sector has been the organization of auctions based on a new model that was established between 2003 and 2004 for long-term planned contracts. This mechanism is based on competition between wind, biomass, hydro and centralized solar (non-distributed) sources, but also with natural gas and coal-based thermoelectric and hydropower plants. However, given the lack of efficiency improvements in existing plants (via retrofit) and investments in new greenfield plants, sugar-cane bioelectricity is still uncompetitive and the highest energy prices (ANEEL[8] 2019). In addition to a lack of differentiation within the biomass category, the auction rules do not take into account the externalities and characteristics of each sugar-cane bioelectricity project (retrofit, greenfield, use of straw and bagasse, biogas production, etc.).

8.5 Is There a Niche for Sugar-Cane Bioelectricity?

Within a regime, niches are the "spaces" where sustainability innovation can take place, and they are influenced by the broader "landscape" and the dynamic of the respective "regime" (Geels 2011). As explained above, sugar-cane bioelectricity is not an innovative practice—and market—per se, but a "niche", since its promotion can stem from meta-discourses.

In the near future, the lack of hydroelectricity is likely to be aggravated by the greater vulnerability associated with extreme events directly related to climate change such as droughts and floods (PBMC[9] 2014). The promotion of sugar-cane bioelectricity, in addition to wind and solar

---

[8] Agência Nacional de Energia Elétrica: Brazilian Electricity Regulatory Agency.
[9] Painel Brasileiro de Mudanças Climáticas: Brazilian Panel on Climate Change.

sources, in principle, would benefit the government and be perfectly in line with meta-discourses on energy transitions.

The first positive feature of bioelectricity is that it has the lowest levels of emissions (in grams of carbon equivalent per kWh) of all forms of renewable energy. It is particularly better than solar or wind energy, which, in addition to being intermittent sources, are currently dependent on natural gas (CTBE[10] 2017). The second positive feature of sugar-cane bioelectricity is its seasonality and complementarity with hydropower, as it is available during the period with the lowest water supply, just as much as it is available during the rainy season. In contrast to wind and photovoltaic sources, which are used for distributed energy generation and are intermittent instead of being dispatchable, biomass provides an uninterrupted production of (bio)electricity (CTBE 2017). Thirdly, the greater insertion of bioelectricity in the electrical system provides a strategic option with which to expand the national grid thanks to the distributed generation of bioelectricity close to the main consumption centres (in the southeast with expansion towards the centre). The inclusion of bioelectricity on a scale commensurate with its potential would reduce the need for investment in strengthening and expanding the electrical grid and would reinforce transmission efficiency by reducing technical losses. Fourthly, the sugar-cane industry can be related to other energy sources such as bioelectricity from straw and tips; biogas mainly from vinasse, which can be used to generate bioelectricity or biomethane, a substitute for natural gas and diesel; and second-generation ethanol from bagasse, straw and tips.

Periods of drought coincide with the months with the highest intensities of wind and therefore peak wind power generation takes placing during these periods (SEBRAE[11] 2017). But they equally coincide with the sugar-cane harvest and thus the possibility of producing bioelectricity (CCEE[12] 2019). In 2019, the supply of sugar-cane bioelectricity to the

---

[10] Laboratório Nacional de Ciência e Tecnologia do Bioetanol: Brazilian Bioethanol Science and Technology National Laboratory.

[11] Serviço Brasileiro de Apoio às Micro e Pequenas Empresas: Brazilian Micro and Small Enterprise Support Service.

[12] Câmara de Comercialização de Energia Elétrica: Chamber of Commercialization of Electric Energy.

grid (21.5 terawatt-hours—TWh) represented a saving of 15% of the total energy stored in the reservoirs of south-eastern/mid-western hydro-electric plants (UNICA 2019). Thanks to the extensive electrical grid that crosses the country, selection pressures should lead to institutional changes promoting the complementarity of the different renewable sources, both regionally, in the form of distributed electricity generation, and nationally, thanks to the advantages of sugar-cane bioelectricity from the point of view of energy security and sustainability.

Based on the regional potential of each source, an expansion of wind and solar energy is expected in the north-eastern and mid-western regions, and sugar-cane bioelectricity is predicted to increase in the southeast, with very little expansion in the south (EPE 2020). However, the technological synergy between gas and biogas (Bradshaw and de Martino Jannuzzi 2019) and short term economic concerns have pushed policymakers in the state of São Paulo (in the southeast), which is responsible for 52% of the sugar-cane bioelectricity generated in Brazil (NovaCana 2020) and home to the largest gas pipeline network in the country, to bet on the development of natural gas and the introduction of a state-sponsored biomass programme. The Brazilian sugar-cane industry can generate 56 million m^3 of biomethane per day, which corresponds to 10,565 megawatt (MW), or 75% of the capacity of the Itaipu hydroelectric plant (NovaCana 2019).

Data from the National Energy Balance (EPE 2017) show that at the end of the 1980s, all of the bioelectricity generated at the time (3.5 TWh in 1987) was destined for self-consumption in the mills. Since 2013, the sugar-energy sector has produced more bioelectricity for the national grid than the volume needed to meet its own electricity demand. Even so, in 2018, only 54% (200 of 369) of the mills in operation commercialized surplus electricity production on the national electricity market; this represents just 15% (21.5 TWh) of the potential provided by sugar-cane bioelectricity (142 TWh) (UNICA 2019). Encouraged by the RenovaBio programme, the number of sugar-cane mills could rise to 390, providing almost 60% more electricity between 2018 and 2030 (ibid.).

Through its dependence on biomass production from the sugar and ethanol markets, bioelectricity has been at the mercy of the swings in the sugar-cane sector that were accentuated after the worldwide crisis in

2008. In terms of the annual evolution of installed capacity, 2010 was a record year for biomass, with 1,750 MW (equivalent to 12.5% of an Itaipu Plant), as the result of investment decisions before 2008, when the expansion of the sugar-energy sector was being promoted (UNICA 2018). Between 2010 and 2011, ethanol production stagnated, investments in new capacity dried up, and the global market proved to be much more modest. The decrease in BNDES disbursements for sugar-cane bioelectricity until the recent RenovaBio programme (NovaCana 2020) can be explained by the retraction of investment in the sugar-cane sector itself, but also by the loss of competitiveness in the regulated auctions promoted by the Brazilian federal government since 2009 (ibid.) (Fig. 8.1).

Although sugar-cane bioelectricity is not an innovative practice or market in itself, it could become a key niche in the energy matrix if the relevant socio-technical transformations that depend on public policies are put in place. In turn, this depends on the degree to which key regime actors are influenced by the global meta-discourses and the way in which they shape global markets and regulation.

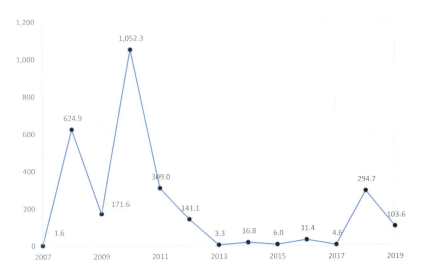

Fig. 8.1 Total BNDES funds contracted by the biomass industry between 2007 and 2019, in million R$ (*Source* NovaCana [2020]. Authors' illustration)

ApexBrasil (Brazilian Trade and Investment Promotion Agency) and UNICA, which represents more than 50% of Brazilian sugar-cane production, joined efforts to promote a "global agenda" of "alternatives to fossil fuels" (Biofuture site). This led the Brazilian government to launch the Biofuture Platform in 2016, a multi-stakeholder initiative with 20 countries, "to accelerate the transition to an advanced, low carbon, global bioeconomy". By using expressions such as "sustainable bioeconomy", "low carbon transport fuels", "renewable energy", "taking advantage of new, sustainable technologies already in place", the platform employs the meta-discourses of energy transition and bioeconomy to foster a global ethanol market, and, thus, the Brazilian sugar-cane industry, as a whole.

The top five bioelectricity generation companies and recipients of BNDES financing in recent years have mainly been Brazilian (NovaCana 2020). First place is occupied by Raízen (the joint venture between the Anglo-Dutch Shell corporation and the Brazilian corporation Cosan). Raízen is one of the promoters of the Brazilian Association of Sustainable Industrial Biotechnology (ABBI), which officially supports the Parliamentary Front of Innovation in Bioeconomy (launched in June 2019), and the Biofuture Platform. This context is favourable to the legitimation of meta-discourses in the negotiation between the actors involved who support institutional changes that encourage the commercialization of sugar-cane bioelectricity.

A series of technical, financial and public policy measures would be necessary to achieve a sugar-cane bioelectricity capacity of 6.7 gigawatt by 2029 (EPE 2020). At the technical level, and in the case of existing plants, new investment is needed to increase the efficiency of energy generation. Furthermore, special credit lines are needed to finance both the new equipment and the costs of connecting to the national grid, which remains the responsibility of bioelectricity producers and is only viable in very favourable locations. At the same time, the political design of the auction system is crucial to the promotion of renewables and, within this category, of bioelectricity. Finally, a long-term and stimulating sector policy for bioelectricity is important, with clear guidelines and continuity with the aim of guaranteeing the full efficient use of this renewable energy resource in the country's energy system.

8.6 Conclusions

The concepts of energy transition and bioeconomy represent meta-discourses (the landscape) that aim to incorporate sustainability concerns into bioenergy governance (the regime), and to incite institutional changes for a new bioenergy market (bioelectricity niche). At present, the complex interactions between the different levels highlights a transition towards an energy mix characterized by natural gas and new renewable energy sources, particularly wind but also sugar-cane bioelectricity and solar. Justifications based on multiple energy security considerations and environmental concerns represent supplementary factors that enhance and legitimize political support for the sugar-cane industry, which, however, still requires a major programme of investments and reformed auction regulations.

The promotion of sugar-cane bioelectricity, in addition to wind and solar sources, makes sense in terms of the energy mix. Thanks to the extensive electrical grid that criss-crosses the country, the sugar-cane industry can reinforce distributed electricity generation, provide bioelectricity to the southeast and south, with wind and solar energies concentrated in the northeast and mid-west, and biogas connected to the largest gas pipeline network in the southeast. Political support for sugar-cane bioelectricity would be in line with energy security, industrial development and sustainability concerns.

Finally, meta-discourses clearly influence the promotion of institutional changes through the development of a global network (Biofuture Platform), new national regulations (RenovaBio), and a new model of electric generation (distributed thanks to wind and solar sources). Both the Platform and the RenovaBio programme illustrate the convergence of interests between areas of the government that are favourable to the international ethanol market and concerned with undertaking action to comply with the Paris Agreement and promoting the transnationalized sugar-cane industry with its interests in expanding its market. However, given the technological obsolescence of bioelectricity production and the critical financial state of the sugar-cane industry in addition to the current priorities of the electricity marketing model, sugar-cane bioelectricity remains uncompetitive and its future uncertain.

References

Andrews-Speed, P. (2016). Applying Institutional Theory to the Low-Carbon Energy Transition. *Energy Research and Social Science, 13*, 216–225.

ANEEL (2019). *Generation Auction: Energy Auction from New Power Generation Projects (A-6 of 2019)*. Brazil.

ANP (2018). Análise do Setor de Gás Natural no Brasil: Medidas para Dinamização do Mercado. Superintendência de Infraestrutura e Movimentação - SIM. http://www.anp.gov.br/images/Palestras/ANP_Firjan_03_12_2018.pdf. Accessed 15 March 2020.

Avelino, F., & Wittmayer, J.M. (2016). Shifting Power Relations in Sustainability Transitions: A Multi-actor Perspective. *Journal of Environmental Policy & Planning, 18*(5), 628–649.

Bernstein, S., & Cashore, B. (2012). Complex Global Governance and Domestic Policies: Four Pathways of Influence. *International Affairs, 88*(3), 585–604.

BP (2019). *BP Energy Outlook 2019 Edition. BP Energy Outlook 2019.* London. https://www.bp.com/content/dam/bp/business-sites/en/global/corporate/pdfs/energy-economics/energy-outlook/bp-energy-outlook-2019.pdf. Accessed 15 March 2020.

Bradshaw, A., & Martino Jannuzzi, G. de (2019). Governing Energy Transitions and Regional Economic Development: Evidence from Three Brazilian States. *Energy Policy, 126*, 1–11.

CCEE (2019). InfoMercado: Dados Individuais. São Paulo, SP, Brazil. https://www.ccee.org.br/portal/faces/pages_publico/o-que-fazemos/infomercado?_adf.ctrl-state=18svg3mlof_5&_afrLoop=78439674075827#!. Accessed 15 March 2020.

CMA (2019). *A Política Nacional sobre Mudança do Clima.* Brasília, DF, Brazil. http://www.mma.gov.br/clima/politica-nacional-sobre-mudanca-do-clima. Accessed 15 March 2020.

Colomer, M., & Queiroz, H. (2019, April 17). Os condicionantes da política energética do setor de petróleo nas últimas décadas. *Blog Infopetro*. https://

infopetro.wordpress.com/2019/04/17/os-condicionantes-da-politica-energe tica-do-setor-de-petroleo-nas-ultimas-decadas/. Accessed 15 March 2020.
Costa, L. (2020, February 20). Solar Energy Expands Quickly in Brazil, Attracts Chinese Firms. *Reuters*. https://www.reuters.com/article/us-brazil-power-solar/solar-energy-expands-quickly-in-brazil-attracts-chinese-firms-idUSKBN20E2UO. Accessed 15 March 2020.
CTBE (2017). Cartilha da Bioeletricidade - Superando as Barreiras Regulatórias para a Comercialização de Eletricidade pelas Usinas do Setor Sucroenergético - Projeto SUCRE. https://www.unica.com.br/wp-con tent/uploads/2019/06/Cartilha-Da-Bioeletricidade.pdf. Accessed 15 March 2020.
Dietz, T., Börner, J., Förster, J.J., & von Braun, J. (2018). Governance of the Bioeconomy: A Global Comparative Study of National Bioeconomy Strategies. *Sustainability (Switzerland)*, *10*(9).
Dubash, N.K., & Florini, A. (2011). Mapping Global Energy Governance. *Global Policy*, *2*(SUPPL.1), 6–18.
EPE (2017). Painel 30 Anos de Bioeletricidade: Realizando o Potencial. In *Ethanol Summit 2017* proceedings. São Paulo/SP. http://www.epe.gov.br/sites-pt/publicacoes-dados-abertos/publicacoes/PublicacoesArquivos/publicacao-244/topico-254/EPE_ETHANOL SUMMIT 2017 _JOSE MAURO_2017_2706.pdf. Accessed 15 March 2020.
EPE (2018). Nota Técnica PR 04/18 - Potencial dos Recursos Energéticos no Horizonte 2050. Rio de Janeiro, Brazil. http://www.epe.gov.br/sites-pt/pub licacoes-dados-abertos/publicacoes/PublicacoesArquivos/publicacao-227/top ico-416/NT04PR_RecursosEnergeticos2050.pdf. Accessed 15 March 2020.
EPE (2019). Balanço Energético Nacional 2019: Relatório Síntese / Ano Base 2018. Rio de Janeiro, Brazil. http://www.epe.gov.br/sites-pt/publicacoes-dados-abertos/publicacoes/PublicacoesArquivos/publicacao-377/topico-470/RelatórioSínteseBEN2019AnoBase2018.pdf. Accessed 15 March 2020.
EPE (2020). Plano Decenal de Expansão de Energia 2029. Brasília, Brazil. http://www.epe.gov.br/sites-pt/publicacoes-dados-abertos/publicacoes/Doc uments/PDE2029.pdf. Accessed 15 March 2020.
EU (2019). Clean Energy for All Europeans. Luxembourg. https://op.europa.eu/en/publication-detail/-/publication/b4e46873-7528-11e9-9f05-01aa75 ed71a1/language-en?WT.mc_id=Searchresult&WT.ria_c=null&WT.ria_f= 3608&WT.ria_ev=search. Accessed 15 March 2020.
Foxon, T.J., & Pearson, P.J.G. (2007). Towards Improved Policy Processes for Promoting Innovation in Renewable Electricity Technologies in the UK. *Energy Policy*, *35*(3), 1539–1550.

Foxon, T.J., Hammond, G., & Pearson, P.J.G. (2010). Developing Transition Pathways for a Low Carbon Electricity System in the UK. *Technological Forecasting and Social Change, 77*(8), 1203–1213.

Garcez, C.G. (2017). Distributed Electricity Generation in Brazil: An Analysis of Policy Context, Design and Impact. *Utilities Policy, 49*, 104–115.

Geels, F.W. (2011). The Multi-level Perspective on Sustainability Transitions: Responses to Seven Criticisms. *Environmental Innovation and Societal Transitions, 1*(1), 24–40.

Geels, F.W., & Schot, J. (2007). Typology of Sociotechnical Transition Pathways. *Research Policy, 36*(3), 399–417.

Goldemberg, J., Schaeffer, R., Szklo, A., & Lucchesi, R. (2014). Oil and Natural Gas Prospects in South America: Can the Petroleum Industry Pave the Way for Renewables in Brazil? *Energy Policy, 64*, 58–70.

Goldthau, A. (2013). *The Handbook of Global Energy Policy*. West Sussex: Wiley-Blackwell.

Grin, J., Rotmans, J., & Schot, J. (2010). *Transitions to Sustainable Development*. New York: Routledge.

Hajer, M.A. (1995). *The Politics of Environmental Discourse: Ecological Modernization and the Policy Process*. New York: Oxford University Press.

Hall, P.A. (2010). Historical Institutionalism in Rationalist and Sociological Perspective. In J. Mahoney & K. Thelen (Eds.), *Explaining Institutional Change: Ambiguity, Agency, and Power* (pp. 204–224). Cambridge: Cambridge University Press.

Hauff, J., Bode, A., Neumann, D., & Haslauer, F. (2014). *Global Energy Transitions: A Comparative Analysis of Key Countries and Implications for the International Energy Debate*. Berlin: Weltenergierat – Deutschland e.V.

Herrera, S. (2014). Análise da Governança Global da Sustentabilidade dos Biocombustíveis e Proposta para o Etanol Brasileiro. Universidade Federal do Rio de Janeiro. http://antigo.ppe.ufrj.br/ppe/production/tesis/selena.pdf. Accessed 15 March 2020.

IEA (2019). Market Report Series: Gas 2019 – Analysis and Forecasts to 2024. https://www.iea.org/reports/market-report-series-gas-2019. Accessed 15 March 2020.

Khallaghi, N., Hanak, D.P., & Manovic, V. (2020). Techno-Economic Evaluation of Near-Zero CO2 Emission Gas-Fired Power Generation Technologies: A Review. *Journal of Natural Gas Science and Engineering, 74*, 103095.

Kuzemko, C., Lockwood, M., Mitchell, C., & Hoggett, R. (2016). Governing for Sustainable Energy System Change: Politics, Contexts and Contingency. *Energy Research and Social Science, 12*, 96–105.

MME (2019a). Resenha Energética Brasileira. Exercício de 2018. http://www.mme.gov.br/documents/36208/948169/Resenha+Energética+Brasileira+-+edição+2019+v3.pdf/92ed2633-e412-d064-6ae1-eefac950168b. Accessed 15 March 2020.

MME (2019b). NOVO MERCADO DE GÁS - Nota Técnica: Propostas Para o Mercado Brasileiro de Gás Natural. Brasília. http://www.mme.gov.br/documents/36112/491930/2.+Relatório+Comitê+de+Promoção+da+Concorrência+vfinal+10jun19.pdf/2379cc7f-f6b7-8ba0-72db-1278e7d252ca. Accessed 15 March 2020.

MRE (2015). Intended Nationally Determined Contribution: Towards Achieving the objective of the United Nations Framework Convention on Climate Change. Intended Nationally Determined Contribution (Vol. 9). http://www.itamaraty.gov.br/images/ed_desenvsust/BRASIL-iNDC-portugues.pdf. Accessed 15 March 2020.

NovaCana (2019, October 29). Mauro Mattoso, do BNDES: "O biogás torna o etanol mais verde ainda." https://www.novacana.com/n/industria/investimento/mauro-mattoso-bndes-biogas-torna-etanol-mais-verde-ainda-201 91029. Accessed 15 March 2020.

NovaCana Database (2020). NovaCana Data. https://www.novacana.com/data/dados/. Accessed 15 March 2020.

Nunes, E. (1997). *A gramática política do Brasil: clientelismo e insulamento burocrático*. Rio de Janeiro: Enap, Jorge Zahar.

Oliveira, A. de (2007). Political Economy of the Brazilian Power Industry Reform. In D.G. Victor & T.C. Heller (Eds.), *The Political Economy of Power Sector Reform* (pp. 31–75). Cambridge: Cambridge University Press.

PBMC (2014). *Impactos, vulnerabilidades e adaptação. Contribuição do Grupo de Trabalho 2 do Painel Brasileiro de Mudanças Climáticas ao Primeiro Relatório da Avaliação Nacional sobre Mudanças Climáticas*. Rio de Janeiro, Brazil. http://www.pbmc.coppe.ufrj.br/documentos/RAN1_completo_vol2.pdf. Accessed 15 March 2020.

Pülzl, H., Kleinschmit, D., & Arts, B. (2014). Bioeconomy - An Emerging Meta-Discourse Affecting Forest Discourses? *Scandinavian Journal of Forest Research, 29*(4), 386–393.

Rip, A., & Kemp, R. (1998). Technological Change. In S. Rayaner & E.L. Malone (Eds.), *Human Choice and Climate Change. Vol. II, Resources and Technology* (pp. 327–399). Columbus: Battelle Press.

Rodríguez-Morales, J.E. (2018). Convergence, Conflict and the Historical Transition of Bioenergy for Transport in Brazil: The Political Economy of Governance and Institutional Change. *Energy Research and Social Science, 44*, 324–335.

SEBRAE (2017). *Cadeia de Valor da Energia Eólica no Brasil*. Brasília, Brazil. https://bibliotecas.sebrae.com.br/chronus/ARQUIVOS_CHRONUS/bds/bds.nsf/1188c835f8e432ddd43bc39d27853478/$File/9960.pdf. Accessed 15 March 2020.

Slaughter, A.-M. (2005). *A New World Order*. Princeton, Oxford: Princeton University Press.

Smith, A., Stirling, A., & Berkhout, F. (2005). The Governance of Sustainable Socio-Technical Transitions. *Research Policy, 34*(10), 1491–1510.

Souza, Z.J. de, & Azevedo, P.F. de (2006). Geração de energia elétrica excedente no setor sucroalcooleiro: um estudo a partir das usinas paulistas. *Revista de Economia e Sociologia Rural, 44*(2), 179–199.

Streeck, W., & Thelen, K. (2005). *Beyond Continuity: Institutional Change in Advanced Political Economies*. Oxford: Oxford University Press.

TCU (2018). TC 008.692/2018-1, Relatório de Auditoria, Grupo I, Classe V. https://portal.tcu.gov.br/data/files/09/12/E4/9A/0052C6105B9484B6F18818A8/008.692-2018-1-AC-matrizeletrica_energiasrenovaveis.pdf. Accessed 15 March 2020.

UNICA (2018). Boletim/UNICA: A Bioeletricidade em números – Setembro/2018. São Paulo/SP, Brazil. https://www.unica.com.br/wp-content/uploads/2019/06/Numeros-da-Bioeletricidade-em-2018-UNICA.pdf. Accessed 15 March 2020.

UNICA (2019). A Bioeletricidade de Cana: Boletim de Julho de 2019. https://www.unica.com.br/wp-content/uploads/2019/07/UNICA-Bioeletricidade-julho2019-1.pdf. Accessed 15 March 2020.

Van Dam, J., Junginger, M., Faaij, A., Jürgens, I., Best, G., & Fritsche, U. (2008). Overview of Recent Developments in Sustainable Biomass Certification. *Biomass and Bioenergy, 32*(8), 749–780.

Wilkinson, J. (2014). Brazil, Biofuels and Bio-Diplomacy with a Specific Focus on Africa and Mozambique. Rio de Janeiro, Brazil. https://www.academia.edu/14738018/Brazil_s_Biofuels_Diplomacy. Accessed 15 March 2020.

Wilkinson, J. (2015). The Brazilian Sugar Alcohol Sector in the Current National and International Conjuncture. ActionAid, Brazil. http://actionaid.org.br/wp-content/files_mf/1493419528completo_sugar_cane_sector_ing.pdf. Accessed 15 March 2020.

Wilkinson, J., & Herrera, S. (2010). Biofuels in Brazil: Debates and Impacts. *Journal of Peasant Studies, 37*(4).

Open Access This chapter is licensed under the terms of the Creative Commons Attribution 4.0 International License (http://creativecommons.org/licenses/by/4.0/), which permits use, sharing, adaptation, distribution and reproduction in any medium or format, as long as you give appropriate credit to the original author(s) and the source, provide a link to the Creative Commons license and indicate if changes were made.

The images or other third party material in this chapter are included in the chapter's Creative Commons license, unless indicated otherwise in a credit line to the material. If material is not included in the chapter's Creative Commons license and your intended use is not permitted by statutory regulation or exceeds the permitted use, you will need to obtain permission directly from the copyright holder.

Part IV

Reconfigurations and Continuities of Social-ecological Inequalities in Rural Areas

9

Buruh Siluman: The Making and Maintaining of Cheap and Disciplined Labour on Oil Palm Plantations in Indonesia

Hariati Sinaga

9.1 Introduction

The oil palm sector is one of the front-runners in the Indonesian agricultural sector. Since 2007, Indonesia has been the largest producer of crude palm oil (CPO) worldwide. The sector is a major source of foreign reserves for the country as well as a main instrument of poverty alleviation and rural economic development (Rist et al. 2010; Zen et al. 2005). While palm oil remains important for Indonesia's food and household goods industries, since 2006, the country has considered utilising palm oil outputs in the transition towards a bioeconomy. The implementation of a bioeconomy in Indonesia is mainly aimed at achieving food security as well as advancing the development of bioenergy (Sudaryanto 2015). The palm oil sector is considered a strategic sector in achieving both of these objectives.

H. Sinaga (✉)
International Center for Development and Decent Work (ICDD), Kassel, Germany

While some studies on palm oil development in the country emphasise its positive impacts, many others shed light on the adverse impact it has on the environment and on people's livelihoods (Richter 2009; Colchester et al. 2006). The working conditions of plantation workers are central to these debates on livelihoods. Reports have documented decent work deficits on Indonesian plantations, which are associated with cheap and disciplined labour as an important feature of the plantation labour regime. This chapter takes a closer look at female labour on plantations.

Although there has been increasing interest in taking gender issues into account in discussions about oil palm plantations in Indonesia, two important things are missing in this debate. Firstly, recent debates have largely concentrated on the gendered impacts of plantation work (Julia and White 2012; Li 2015; Elmhirst et al. 2017). Although this is justifiable, as scholars have just begun to take up gender as an analytical framework in assessing plantation works, we should go further and use a feminist lens to trace and understand the production of the social subject, namely plantation labour. Secondly, some studies of oil palm plantations in Indonesia discuss the labour regimes that have been established on plantations, and this includes the gender regime, but they do not sufficiently employ historical analyses of the production of such labour regimes, especially those concerning female plantation workers (Li 2015, 2017). This chapter addresses this research gap. Contributing to the discussions on female labour on oil palm plantations in Indonesia, it seeks to provide a historical analysis of the construction of female labour on plantations, which, until now, has remained underexplored. Drawing on insights from feminist theories, coloniality/modernity scholarship, as well as literature on racial capitalism, this chapter argues that female labour on plantations, often called *buruh siluman*, plays a central role in and maintaining labour relations that rely on cheap and disciplined labour. The next section is structured as follows: I start by drafting a theoretical framework with which to examine the changing role of female labour on Indonesian palm oil plantations. I then discuss the historical development of the plantation labour regime and focus on the (re-)production of women as a specific plantation labour subject. Afterwards, I examine the current working conditions of women on oil palm plantations in Riau, a province that hosts the largest oil palm plantations in Indonesia. I conclude by discussing the role played by women in the making and maintaining of cheap and disciplined labour on oil palm plantations.

9.2 Moving Beyond Working Conditions: Theoretical Remarks

Working conditions emphasise the labour regime installed in a workplace. As a concept, "labour regime" describes a terrain of struggles between capital and labour that are mediated by the state (Cumbers et al. 2008, p. 373; Selwyn 2011). Among others, discussions on labour regimes enable us to scrutinise how a plantation labour subject is constructed (Coe and Jordhus-Lier 2010, p. 13) in order to shed light on reciprocities between labour regime and labour agency (Rodriguez and Mearns 2012). This chapter focuses on the macro-labour control regime, which concerns with capitalist relations of production (Pattenden 2016).

Whereas orthodox Marxist analyses treat "primitive accumulation" as the foundation of the rise in capitalist wage-labour relations, this process is ongoing, especially in the Global South (Fairbairn et al. 2014). Although the rise in wage labour, which is a feature of advanced capitalist society, also occurs in the Global South, a large number of forms of non-wage or informal labour remain. Building on Anibal Quijano's (2000) "coloniality of power", Manuela Boatcă (2013) introduces the term "coloniality of labour" in order to describe co-existing modes of labour control. Furthermore, expanding Marx's analysis, Cedric Robinson (1983) argues that capitalism did not break from the feudal order, but rather evolved from it to produce a "racial capitalism" that depends on slavery, violence, imperialism, and genocide.

Feminist critique of Marx's analysis on primitive accumulation points at the absence of female subjugation as an important aspect in this process (Federici 2014). Maria-Rosa Dalla Costa (1971) argues that women assigned the role of housewives are important in producing the commodity of labour power. The nuclear family, thus, is a social factory where this commodity is produced and women as housewives are disciplined. The female subjugation process is also shown by Maria

Mies' (1994) work on the process of "housewifisation" as an important element in the history of capitalist development. Drawing on world-systems theory, Mies connects the processes in the West and those in the colonies to form a systematic and historical process that involves the exploitation of women, nature, and the colonies and that shapes the sexual division of labour both in the West and in the colonies. This is in line with Maria Lugones' (2007) concept of the "coloniality of gender", in which colonisation serves as a gendered act and thus intensifies the gender hierarchies in colonised societies. Mies' work also informs much of feminist subsistence theories (Mies and Bennholdt-Thomsen 2000), which show that: (1) the production of large marginalised masses in the periphery was integral to the capitalist mode of production; (2) this development was based on the economic position of the housewife both in the centres and in the peripheries in the capitalist world-economy.

To sum up, these theoretical insights provide lenses with which to understand the making and maintaining of the plantation labour subject. First, the (re-)production of social differences and sameness are important in capitalist development. Second, primitive accumulation is an ongoing process that is not only manifested in the persistent occurrence of land dispossession, but also in the enduring process of female subjugation. As I will show, the persistence of land dispossession and the process of female subjugation are interwoven processes that constitute female labour on oil palm plantations. In the following, I provide a historical overview of labour relations on oil palm plantations in order to illustrate the construction of women plantation workers, *buruh siluman*, and how this plantation subject is important in sustaining a cheap and disciplined labour regime.

9.3 Women "Coolies", *Nyai*, and the (Re-)Production of a Plantation Labour Subject

Oil palm plantations in Indonesia started in Sumatra in 1911 (Manggabarani 2009), particularly in Deli. Before oil palm, tobacco, coffee, and rubber were cultivated in Deli. The creation of the Deli plantation society involved the recruitment of "coolies" from Java, Penang, and other places such as China because the local inhabitants were unwilling to work on plantations as wage workers (Breman 1989). "Coolie" refers to an indentured service that aims to provide cheap and well-controlled labour. Its use followed the dismantling of slavery, and the establishment of a modern racial governmentality of "free" yet racialised and coerced labour (Lowe 2015, p. 24). This contradiction is evident in the practice of indenture labour on plantations in Deli, as its use followed the dismantling of the cultivation system imposed by the Dutch colonial authorities as well as the shift towards an open-door policy.

While women "coolies" were a minority on Sumatran plantations at the time and accounted for a merely 8% of workers in the beginning of the twentieth century, their numbers gradually increased as oil palm plantations expanded and had risen to one quarter of the total workforce by 1930 (Breman 1989, p. 95). These women were mainly young Javanese as they were considered docile. Women "coolies" worked on plantations as well as in the social reproduction sphere, such as prostitution. The increasing prevalence of Javanese women "coolies" was the result of various factors. The gradual phasing out of the concubinage[1] between European men and native, mainly Javanese, women in the Dutch East Indies, the name for Indonesia at the time, contributed to an increase in prostitution among native women (Ming 1983, as cited in Ingleson 2013, p. 215). As mentioned above, the dismantling of the cultivation system, a system that resulted in the declining welfare of Javanese populations (van Nederveen Meerkerk 2017, p. 40), paved the way for the shift from indentured labour to free labour in Java. Women who could not find work on plantations or factories entered into prostitution. The open-door policies embraced by the Dutch colonial authorities attracted foreign, mainly European, investors. They opened up plantations, such as the Deli plantations, as well as sugar factories in sparsely populated areas of Java or the Outer Islands. As we can see from the case of plantations in Deli, male workers had to be sourced from

---

[1] During the early Dutch colonisation of the Dutch East Indies, only Dutch men were allowed to settle in the colony. This paved the way for the use of native, mainly Javanese, women as sex slaves. It was only when European women arrived in the colony in large numbers in the late nineteenth century that concubinage was gradually phased out.

elsewhere, mainly from densely populated areas of Java. This then led to a rise in prostitution in the newly opened plantations and factories.

According to Breman (1989), planters had become more favourable to Javanese women "coolies" as new coffee plantations in Serdang shifted towards piece rate. The piece rate enabled planters to accumulate more profits as they discovered that women "coolies" were the cheapest labour on plantations. In general, the gradual increase in number of women "coolies" was in line with the low wage strategy, which gained importance on plantations during the twentieth century (p. 109).

Apart from women "coolies", the social reproduction sphere on plantations was also the responsibility of *nyai*, which refers to native (mainly Javanese) women who were either concubines or housekeepers. The involvement of native women as concubines might not be straightforward and could be disguised under the colonial euphemism of "housekeeper", especially after concubinage had been phased out. While the presence of *nyai* was more prevalent in Dutch colonial settlements in Java, their existence saw increasing numbers in the Outer Islands where the Dutch found their footings. On plantations in Deli, marriage was prohibited for incoming European employees, as planters believed that employees with families would not be able to support themselves properly (Stoler 2010, p. 29). Hence, employing *nyai* was seen as a convenient arrangement.

In her elaboration of the "housewifisation" process, Maria Mies (1994) highlights the disruption of subsistence in the colonies. This is evident in the case of Java, where many women "coolies" came from. One of the logics behind the Dutch cultivation system in Java was to teach "lazy" indigenous peasants, who practised subsistence, the virtue of industriousness (van Nederveen Meerkerk 2017, p. 39). The cultivation system disrupted the subsistence system of Javanese peasants, which subsequently had an impact on labour relations (ibid., p. 42). The system, which catered to commercial plantations, contributed to the loss of women's prerogatives under subsistence agriculture.[2] Elise van Nederveen Meerkerk (2017) connects the changes in labour relations under the

---

[2] The typical egalitarian rice-growing Javanese societies viewed rice-growing activities on dry lands and garden agriculture as a women's prerogative, while both women and men cultivated rice on wetland (van Nederveen Meerkerk 2017, p. 41). The cultivation system forced the male

cultivation system to those in the metropole. The colonial gains made it possible to raise male wages in the Netherlands, which paved the way for the formation of typical bourgeois family ideals, and married women in the workforce became increasingly rare. This process, which Mies argues led to the emergence of Dutch housewives (1994, p. 96), was linked to the disruption of families and homes among estate workers in the Dutch colonies as subsistence agriculture was suppressed. The process that created women "coolies" and *nyai* also involves what Maria Lugones (2007) calls the "coloniality of gender" as it was brought about during colonial encounters. This process was fully realised when plantation workers started to be recruited as families, as discussed in the following.

The increasing criticism of the Coolie Ordinance,[3] the labour regime, and the situation of "coolies" on plantations in Sumatra put pressure on European planters. Formal indenture was phased out after 1910, yet the labour relations that followed were only slightly less coercive (Li 2017). The prevailing characteristic labour regime was gradually replaced by the increasing employment of Javanese married workers, which was known as the "family formation" approach (Stoler 2010, p. 31). This led to changes in the plantation labour regime. "Coolie" barracks were replaced by dwellings for individual families or by labour compounds with subsistence plots, which resembled village life. Such plots, however, were insufficient to meet subsistence needs. As Stoler puts it, "nominal land allotments represented both a rationale for depressed wages and a relatively cheap means of providing the semblance of village life" (2011, p. 40). Furthermore, Stoler argues that the semblance of village life, particularly village life in Java, was an important labour control, especially during the shift from indentured labour to free labour on the East Sumatra plantations (ibid., p. 38). The "family formation" approach marks the shift from coercive labour control to the kind that relies on

---

peasants out of subsistence agriculture, leaving women and children mostly to undertake this activity.
[3]The Coolie Ordinance came into place in 1888. It regulated employment practices on the plantations during the colonial period; it is deemed more coercive as the regulation was perceived as legitimising the exploitation that occurred in the earlier period (Said 1977, p. 69).

disciplinary power. Stephanie Barral (2014) describes the latter as paternalistic labour relations that exercise strict control over workers' working and private lives. Thereby the "family formation" approach obscures depressed wages. By allowing workers to bring their families into plantations, planters obtained additional labourers. These additional labourers consisted mainly of workers' wives. In other words, the "family formation" approach changed the ways in which women were involved in plantation labour. Recalling the feminist critique of the nuclear family as a site where women are disciplined, and the perception of Javanese women as docile, women were (re-)produced as disciplined plantation labour.

Despite the different labour regimes, women continue to shoulder the burden of productive and reproductive work. Their double role on plantations become the *raison d'etre* for employing women as casual and unpaid labourers, who are subsequently considered cheap labour on plantations. This is evident in the following case study of plantations in Riau, where I discuss the current role of women workers and their working conditions.

9.4 Working Conditions of Female Labour on Oil Palm Plantations in Riau

Riau is an Indonesian province with the largest oil palm plantations in the country. In 2010, oil palm plantations covered 2 million hectares in the province and produced almost 30% of Indonesia's CPO output (Directorate General of Estate Crops 2011, p. 9). As part of my doctoral thesis (Sinaga 2020), I conducted a field research in three company-operated plantations as well as smallholder-owned plantations in Riau in April 2012. One of the companies is a parastatal company, while the rest are private plantation companies. The latter are subsidiaries of two foreign-owned groups considered "big" players in the oil palm sector in Indonesia and Malaysia. Both of these groups operate a substantial number of oil palm plantations in Indonesia. I interviewed 21 workers aged between their mid-20s and mid-50s, twelve of whom were women.

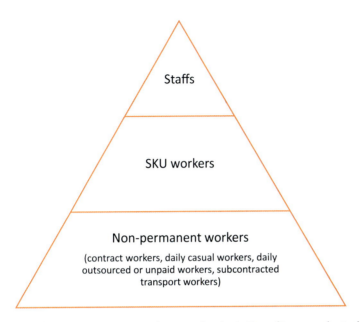

Fig. 9.1 Employment structure on large-scale plantations (*Source* adapted from Siagian et al. 2011, p. 5)

It is difficult to find exact figures for the number of workers employed on oil palm plantations in Indonesia given the rampant practices of employing casual labour.[4] As Fig. 9.1 shows, the employment structure on large-scale oil palm plantations can generally be depicted as a pyramid with the staff at the top and plantation workers on the lower ranks (Siagian et al. 2011, p. 5). Plantation workers are divided into "SKU" (*Syarat Kerja Umum*: general work requirement)[5] workers, non-permanent workers and unpaid labourers, with the latter two at the bottom of the pyramid. There are four types of non-permanent workers (Assalam and Parsaoran 2018). The first type refers to contract workers,

---

[4] Estimates suggest that around 21 million people are employed both directly and indirectly by Indonesia's oil palm plantations (Indonesia Chamber of Commerce and Industry, cited in Mongabay 2019). Sawit Watch estimates that 70% of the workers on plantations are day labourers (Koalisi Buruh Sawit Indonesia 2018).

[5] Although workers in this category are also considered permanent workers, they do not always have contracts and/or receive payslips.

who work for a period of two years in accordance with the national labour law. Contract employment applies to harvesting activities. The second type is daily casual workers, who are mainly responsible for maintenance activities. The third type is daily outsourced workers, which refers to assistants employed to collect loose fruit. The fourth type is subcontracted transport workers.

The structure of employment and classification of workers discussed above are confirmed on the plantations operated by private and parastatal companies visited in Riau.

As it is the case on most oil palm plantations in Indonesia, the women on plantations that I visited work as either paid or unpaid workers. There is a gendered division of labour on oil palm plantations: harvesting is male-dominated as it is considered physically demanding, while maintenance is female-dominated. As mentioned above, most of the maintenance activities are carried out by casual workers. Women are only hired for maintenance work under SKU contracts on one estate managed by a private plantation company. The overseers, however, are (fore)men. The companies claim that female labourers lack leadership capabilities. On the other estate managed by a parastatal company, female workers undertaking maintenance activities are the wives of the harvesters responsible in that area. Another type of work carried out by women workers on plantations is daycare work, assuming daycare facilities exist on the plantations. As unpaid workers,[6] women assist their harvester husbands to collect loose fruit. On plantations owned by smallholders, women workers predominantly serve as unpaid workers helping their husbands.

With regard to wages, SKU workers receive a basic salary and premium payment. Each SKU worker has a daily target to meet. If workers exceed their target, they receive a bonus—an additional payment alongside their basic salary. Loose fruit collected are calculated separately. This system triggers the employment of assistants or unpaid workers. Harvesters clearly desire to get the highest possible premium payment. As such, they employ assistants when daily targets increase, especially

---

[6] Bearing in mind the difficulty of estimating the number of casual labourers on Indonesian oil palm plantations, it is even more difficult to calculate the exact number of unpaid women workers on plantations.

during peak seasons. While assistants can be their relatives or friends, harvesters usually bring their wives and/or children. One SKU harvester stated that when his wife does not help him, his daily yield drops by as much as 50% (male plantation worker, Interview no. 1). This shows the importance of women working as unpaid workers on the plantations.

Women in paid work are mainly casual workers, which means that they do not receive regular income. On one of the plantations operated by a private company, women casual workers not only lack a regular income, but also struggle to gain enough working days in a month in order to earn a living wage. Furthermore, on the other plantation operated by a private company, women who work in maintenance are employed as SKU workers but are still paid less than the minimum wage.[7] Besides facing the issues related to minimum wage,[8] women SKU workers are also disproportionately affected by the fact that the indicators of decent living needs used to set minimum wage levels are gender-biased and thus do not take into account women's decent living needs. As unpaid workers, the income of female workers is tied to the income of their husbands. As mentioned above, these workers actually play a significant role in increasing their husbands' income. This confirms the argument of feminist theories discussed earlier on how female subjugation is important in capitalist relations of production.

On the issue of working hours, workers commonly start working at 7 am. The wives of harvesters arrive at the same time as their husbands, or a bit later. In some cases, they finish working at the same time as their husbands (usually 5 pm on company-operated plantations and 1 pm on plantations owned by smallholders) or earlier. The payment system (e.g. basic salary and premium) encourages harvesters to work overtime. Without helpers or assistants, harvesters would have to work longer on the plantations to meet their targets. The working hours of female maintenance workers are shorter than those of harvesters. Depending on the

---

[7] On estate Y, SKU workers receive Rp 1,133,500 + *premi*. The stipulated minimum wage amounts to Rp 1,389,450.

[8] The indicators of decent living needs are drawn up by the central government under Permenakertrans No. 17/2005. The regulation lists 46 items serving as the basis for a decent living needs survey at the regional level. Workers have demanded that the government revise the regulation to include 122 items.

kind of maintenance activities they undertake and on the company's regulations, they may be provided with a break. Although the length of working hours seems quite modest, in the sense that there is no excessive overtime (as might be the case with factory workers), it is worth remembering that plantation work is physically demanding.

Furthermore, women workers carry a double burden: they work on the plantations as well as at home.[9] At home, they are responsible for reproductive activities. As such, every day, working women have to wake up earlier and go to bed later than their husbands. Most of the women workers I spoke to only cook once a day because they do not have the time or energy to do so more often.

9.5 Cheap and Disciplined Labour as a Key Feature of Labour Relations on Oil Palm Plantations

The discussion about the situation of female workers on plantations in Riau reveal at least four key features female workers are facing today. First, women working as paid casual workers are illustrative of the precarious situation that these women face in terms of irregular income. This situation is exacerbated when women work for too few days in a month to sustain a living wage. Second, while some women workers are employed as permanent workers, their wages are below the stipulated minimum wage. Also, the gender-biased decent living indicators used to determine minimum wage disproportionately affect women. Third, women working as unpaid labourers face income dependency despite the significant contribution they make to the income of their spouses. Fourth, women bear the brunt of oil palm plantation developments, which is demonstrated by the double burden carried by women workers.

I argue that tracing the historical construction of plantation labour subject is important in order to understand the above findings. Tania M. Li (2011, p. 288) argues that cheap, abundant, and disciplined labour

---

[9]A study by Surambo et al. (2011) also shows the double burden of work carried by women workers on oil palm plantations in Central Sulawesi province.

acts as a significant backbone for profit-making on oil palm plantations in Indonesia, whether or not they use contract farming. Citing Ann Stoler's (2011) important work on Indonesia's plantations in Sumatra, Li argues that labour reserves achieved through transmigration programmes found their inception during the colonial period when planters and colonial authorities were debating whether to recruit family or individual labour. I propose to extend the historical analysis to the labour relations on plantations in Sumatra during the colonial period. As discussed earlier, the shift in labour relations on oil palm plantations marks the change from coercive to disciplinary power. Nonetheless, cheap and disciplined labour remains an essential feature of labour relations on plantations. As I argue here, women play a vital role in the making and maintaining of cheap and disciplined labour.

As discussed earlier, ongoing primitive accumulation involves interwoven processes of land dispossession and female subjugation. In the case of Indonesian oil palm plantations, this is demonstrated by the process of land dispossession, which simultaneously restructures gender relations, and which more or less resembles the shift in gender relations that followed the implementation of the cultivation system discussed earlier. Since this chapter does not focus on land dispossession, I draw on recent studies of this issue (Julia and White 2012, p. 1002; Li 2015; Elmhirst et al. 2016) in order to provide a better picture of these interwoven processes. These studies show how this process expropriates lands from indigenous and local people as well as the resulting shift in gender relations, mirroring the "family formation" approach. The latter refers to a gendered land tenure in which the husband is considered the household's head, providing men greater access to income as well as in decisions to sell or mortgage. This practice is essentially used in the various contract farming schemes.[10] As land dispossession also makes subsistence farming obsolete (Federici 2019, p. 77), some of the local or indigenous people may be able to keep a part of their subsistence plots, while the rest is converted to oil palm plantations. Indeed, as it is argued, the remaining importance of subsistence principles among indigenous people is an

---

[10] Under the New Order period, during which the initial oil palm development in the post-independence era was pursued, the state assigned gendered roles to women as wives and mothers, known as "state ibuism" (Suryakusuma 2011).

outcome of plantation companies' efforts to recruit labour from elsewhere (Elmhirst et al. 2016). Companies are able to sustain a cheap and disciplined labour force through depressed wages and ethnic diversification. Women are mainly responsible for these remaining subsistence plots. Nonetheless, women are not only responsible for reproduction, but also for work on the oil palm plantation plots (Julia and White 2012, p. 1003) in order to sustain a minimum living wage. This demonstrates the arguments of feminist subsistence thinkers on the importance of non-wage labour for capitalist accumulation. The consequence of this process is that female labour becomes intensified. The gender dimensions of oil palm development, therefore, simultaneously reverse gender relations from more or less equal to unequal relations with women carrying more of the burden.

Findings from my case study on oil palm plantations in Riau are in line with other studies of working conditions facing women workers on oil palm plantations in Indonesia (Assalam and Parsaoran 2018; Sawit Watch 2017; Li 2015). The poor working conditions facing women on oil palm plantations are generally associated with their presence on plantations as invisible labour, so-called *buruh siluman*. Some argue that these poor working conditions are the result of oil palm developments, while others contend that these situations actually show how cheap and disciplined labour is intrinsic to oil palm developments (Li 2011). As I have shown in this chapter, tracing the historical construction of women as plantation labour subject sheds light on processes through which women as cheap and disciplined labour are (re-)produced on the plantations. Rather than viewing the decent work deficit on plantations as a negative impact of oil palm development, this chapter argues that labour relations that rely on cheap and disciplined labour constitute the capitalist development of the oil palm plantation sector in Indonesia. As the palm oil sector is increasingly regarded as a strategic sector in the country's transition towards bioeconomy, the social inequality in terms of the exploitation of women as cheap and disciplined labour contextualises and is being reproduced in the development of the sector.

List of Interviews quoted

Interview no.	Gender and job position	Business type	Date and place
Interview no. 1	Male plantation worker	Parastatal estate	07/04/2012, Riau

References

Assalam, R., & Parsaoran, H.Z. (2018). Keuntungan di atas manusia: kondisi kerja di bawah rantai pasokan perkebunan sawit milik sinar mas. *Asia Monitor Research Centre and Sawit Watch*.

Barral, S. (2014). Labour Issues in Indonesian Plantations, from Indenture to Entrepreneurship. *Global Labour Column, 177.* http://www.global-labour-university.org/fileadmin/GLU_Column/papers/no_177_Barral.pdf. Accessed 18 Nov 2020.

Boatcă, M. (2013). Coloniality of Labor in the Global Periphery: Latin America and Eastern Europe in the World System. *Review, 36*(3–4), 287–314.

Breman, J. (1989). *Taming the Coolie Beast: Plantation Society and the Colonial Order in Southeast Asia.* New York: Oxford University Press.

Coe, N.M., & Jordhus-Lier, D.C. (2010). Constrained Agency? Re-evaluating the Geographies of Labour. *Progress in Human Geography, 35*(2), 211–233.

Colchester, M., Jiwan, N., Andiko, S.M., Firdaus, A.Y., Surambo, A., & Pane, H. (2006). *Promised Land: Palm Oil and Land Acquisition in Indonesia-Implications for Local Communities and Indigenous People.* Forest People Programme and Perkumpulan Sawit Watch. https://www.forestpeoples.org/sites/default/files/publication/2010/08/promisedlandeng.pdf. Accessed 18 Nov 2020.

Costa, M-R.D. (1971). Women and the Subversion of the Community. In M.-R.D. Costa & S. James (Eds.), *The Power of Women and the Subversion of the Community* (pp. 25–26). Bristol: Falling Wall Press.

Cumbers, A., Nativel, C., & Routledge, P. (2008). Labour Agency and Union Positionalities in Global Production Networks. *Journal of Economic Geography, 8*(3), 369–387.

Directorate General of Estate Crops. (2011). *Tree Crop Estate Statistics of Indonesia*. Jakarta: Directorate General of Estate Crops.

Elmhirst, R., Siscawati, M., Basnett, B.S., & Ekowati, D. (2017). Gender and Generation in Engagements with Oil Palm in East Kalimantan, Indonesia: Insights from Feminist Political Ecology. *The Journal of Peasant Studies, 44*(6), 1135–1157.

Elmhirst, R., Siscawati, M., & Colfer, C.J.P. (2016). Revisiting Gender and Forestry in Long Segar, East Kalimantan, Indonesia: Oil Palm and Divided Aspirations. In C.J.P. Colfer, B.S. Basnett, & M. Elias (Eds.), *Gender and Forests: Climate Change, Tenure, Value Chains and Emerging Issues* (pp. 300–318). New York: Routledge.

Fairbairn, M., Fox, J., Isakson, S.R., Levien, M., Peluso, N., Razavi, S., Scoones, I., & Sivaramakrisnan, K. (2014). Introduction: New Directions in Agrarian Political Economy. *The Journal of Peasant Studies, 41*(5), 653–666.

Federici, S. (2014). *Caliban and the Witch: Women, the Body, and Primitive Accumulation*. New York: Autonomedia.

Federici, S. (2019). *Re-enchanting the World: Feminism and the Politics of the Commons*. Oakland: PM Press.

Ingleson, J. (2013). *Perkotaan, Masalah Sosial dan Perburuhan di Jawa Masa Kolonial*. Depok: Komunitas Bambu.

Julia & White, B. (2012). Gendered Experiences of Dispossession: Oil Palm Expansion in a Dayak Hibun Community in West Kalimantan. *The Journal of Peasant Studies, 39*(3–4), 995–1016.

Koalisi Buruh Sawit Indonesia. (2018). Lembar Fakta Perlindungan Buruh Sawit Indonesia. http://www.turc.or.id/wp-content/uploads/2018/07/Lembar-Fakta-Koalisi-Buruh-Sawit-Indonesia-2018.pdf. Accessed 1 March 2020.

Li, T.M. (2011). Centering Labor in the Land Grab Debate. *The Journal of Peasant Studies, 38*(2), 281–298.

Li, T.M. (2015). Social Impacts of Oil Palm in Indonesia: A Gendered Perspective from West Kalimantan. CIFOR Occasional Paper 124.

Li, T.M. (2017). The Price of Un/Freedom: Indonesia's Colonial and Contemporary Plantation Regimes. *Comparative Studies in Society and History, 59*(2), 245–276.

Lowe, L. (2015). *The Intimacies of Four Continents*. Durham: Duke University Press.

Lugones, M. (2007). Heterosexualism and the Colonial/Modern Gender System. *Hypatia, 22*(1), 186–209.

Manggabarani, A. (2009). *Palm Oil: A Golden Gift from Indonesia to the World*. Directorate General of Estate Crops in Collaboration with Sinar Mas.

Mies, M. (1994). *Patriarchy and Accumulation on A World Scale: Women in the International Division of Labour*. London: Zed Books.

Mies, M., & Bennholdt-Thomsen, V. (2000). *The Subsistence Perspective: Beyond the Globalised Economy*. London: Zed Books.

Mongabay. (2019). Menyoal Nasib Buruh Perkebunan Sawit di Indonesia. https://www.mongabay.co.id/2019/05/01/menyoal-nasib-buruh-perkebunan-sawit-di-indonesia/. Accessed 1 March 2020.

Pattenden, J. (2016). Working at the Margins of Global Production Networks: Local Labour Control Regimes and Rural-Based Labourers in South India. *Third World Quarterly, 37*(10), 1809–1833.

Quijano, A. (2000). Coloniality of Power, Eurocentrism, and Latin America. *Nepantla: Views from South, 1*(3), 533–580.

Richter, B. (2009). Environmental Challenges and the Controversy About Palm Oil Production—Case Studies from Malaysia, Indonesia and Myanmar. Friedrich-Ebert-Stiftung Singapore. http://library.fes.de/pdf-files/iez/06769.pdf. Accessed 20 Oct 2020.

Rist, L., Feintrenie, L., & Levang, P. (2010). The Livelihood Impacts of Oil Palm: Smallholders in Indonesia. *Biodiversity and Conservation, 19*(4), 1009–1024.

Robinson, C.J. (1983). *Black Marxism: The Making of Black Radical Tradition*. London: The University of North Carolina Press.

Rodriguez, J.K., & Mearns, L. (2012). Problematising the Interplay Between Employment Relations, Migration and Mobility. *Employee Relations, 34*(6), 580–593.

Said, M. (1977). *Koeli Kontrak Tempo Doeloe: Dengan Derita dan Kemarahannya*. Medan: Percetakan Waspada.

Sawit Watch. (2017). Ketika Perkebunan Sawit Merampas Kehidupan Perempuan. https://sawitwatch.or.id/2017/03/ketika-perkebunan-sawit-merampas-kehidupan-perempuan-part-1-pendahuluan/. Accessed 1 March 2020.

Selwyn, B. (2011). The Political Economy of Class Compromise: Trade Unions, Capital–Labour Relations and Development in North East Brazil. *Antipode, 43*(4), 1305–1329.

Siagian, S., Hotler, A., Benhidris, S., Jiwan, N., & Hasibuan, F. (2011). Miljoenen van Palm Olie Plantages: A Working Paper on Labour Rights Situation and A Recommendation to Build Just Labour System Guideline for Palm Oil Plantation Workers in Indonesia. Kelompok Pelita Sejahtera, Lentera Rakyat and Sawit Watch.

Sinaga, H. (2020). *Competitive Pressures and Labour Rights: The Indonesian Oil Palm Plantation and Automobile Sectors*. Augsburg: Rainer Hampp Verlag.

Stoler, A.L. (2010). *Carnal Knowledge and Imperial Power: Race and the Intimate in Colonial Rule*. Los Angeles: University of California Press.

Stoler, A.L. (2011). *Capitalism and Confrontation in Sumatra's Plantation Belt, 1870–1979*. Ann Arbor: Michigan University Press.

Sudaryanto, T. (2015). Grand Strategy of Agricultural Development 2015–2045: Sustainable Agricultural Bioindustry as Solution to Future Development in Indonesia. https://ap.fftc.org.tw/article/902. Accessed 19 Nov 2020.

Surambo, A., Susanti, E., Herdianti, E., Hasibuan, F., Fatinaware, I., Safira, M., et al. (2011). Sistem Perkebunan Kelapa Sawit Memperlemah Posisi Perempuan. Laporan Penelitian Sawit Watch dan Solidaritas Perempuan.

Suryakusuma, J.I. (2011). *Ibuisme Negara: konstruksi sosial keperempuanan Orde Baru*. Depok: Komunitas Bambu.

Van Nederveen Meerkerk, E. (2017). Entangled Histories: Unravelling the Impact of Colonial Connections of Both Javanese and Dutch Women's Work and Household Labour Relations, c. 1830–1940. *Tijdschrift Voor Genderstudies, 20*(1), 35–59.

Zen, Z., Barlow, C., & Gondowarsito, R. (2005). Oil Palm in Indonesian Socio-Economic Improvement: A Review of Options. Australia National University Departmental Working Paper No. 2005-11. https://ccep.crawford.anu.edu.au/acde/publications/publish/papers/wp2005/wp-econ-2005-11.pdf. Accessed 19 Nov 2020.

Open Access This chapter is licensed under the terms of the Creative Commons Attribution 4.0 International License (http://creativecommons.org/licenses/by/4.0/), which permits use, sharing, adaptation, distribution and reproduction in any medium or format, as long as you give appropriate credit to the original author(s) and the source, provide a link to the Creative Commons license and indicate if changes were made.

The images or other third party material in this chapter are included in the chapter's Creative Commons license, unless indicated otherwise in a credit line to the material. If material is not included in the chapter's Creative Commons license and your intended use is not permitted by statutory regulation or exceeds the permitted use, you will need to obtain permission directly from the copyright holder.

10

Superexploitation in Bio-based Industries: The Case of Oil Palm and Labour Migration in Malaysia

Janina Puder

10.1 Introduction: Bioeconomy as Green Capitalism

How we investigate social inequalities in an evolving bioeconomy depends on our perception of it and on the sphere we focus on while attempting to grasp the dynamic developments within relevant social relations. From a politico-economic perspective, bioeconomy can be seen as an attempt to reconfigure patterns of production, consumption and circulation (OECD 2019). Although different bioeconomy visions share the goal of establishing a socially and environmentally sustainable economy (Backhouse et al. 2017), none of them questions the fact that bioeconomy is ultimately built on the prevailing principles of capitalism (Goven and Pavone 2015). Consequently, any state policy striving for a bio-based transformation of the economy plays by the common rules

J. Puder (✉)
Friedrich-Schiller-Universität Jena, Jena, Germany
e-mail: janina.puder@uni-jena.de

of capital accumulation (including free market competition, growth and de-/commodification) and the exploitation of labour.

The political actors advocating bioeconomy push for the substitution of fossil energies with green, renewable energy sources. However, this would require a vast increase in the production of biomass (Scarlat et al. 2015). Palm oil is currently one of the most competitive vegetable oils on the global market. It is a versatile crop with a high energy density, and the industry still has a large capacity for growth (Choong et al. 2018). As such, it is a favoured source of biomass for the global bioeconomy.

With a market share of more than 33% in 2018, Malaysia is the second largest palm oil producer in the world after Indonesia (Statista 2019). Between 1995 and 2016, the total area planted with oil palm in Malaysia more than doubled from 2,54 to 5,74 million hectares (Ismail 2013). In 2017, more than 73% of agricultural land in Malaysia was covered with oil palms (Kotecha 2018, p. 2). In recent years, Malaysian businesses have started developing more land in neighbouring Indonesia (Varkkey 2013). The figures stated here are not only important for attempts to retrace the expansion of oil palm in the region but also to gain an understanding of the conditions in which a growing number of rural workers work and live.

In Southeast Asia, palm oil is often associated with social inequalities concerning land ownership, land use and access to land (Li 2009; Pichler 2015) as well as with environmental degradation (Obidzinski et al. 2012; Wakker 2005). The exploitation of migrant workers is a further significant, albeit lesser-known, expression of social inequality that has been caused by industrial oil palm cultivation and the steady expansion of the palm oil sector in Malaysia since the 1960s. With estimations suggesting that more than one million foreign workers are employed in the palm oil plantation and mill sector (Pye et al. 2016), 'low-skilled' migrant workers represent the largest group of workers in the industry (Ismail 2013, pp. 19–20). Investigating the specific working and living conditions of this group, therefore, is crucial to examining existing, solidifying or evolving social inequalities in emerging bio-based industries. Most Malaysian policies that target poverty reduction to close the income gap between the rural and the urban population disregard the importance of migrant labour for the overall performance of the economy, job creation and average income development (World Bank 2015, p. 2). In doing so, the state only addresses the tip of the iceberg in terms of social inequalities in rural areas and neglects the often poor working and living conditions faced by migrant workers (Puder 2019).

Academic literature on green economy models often underexposes the possible effects of a green transformation on existing labour relations (Anderson 2016; Birch 2019; Brand and Wissen 2015). Scenarios discussed by the OECD (2017) or the German Federal Environment Agency (UBA 2014) hint that promoting green industry branches in industrialized countries could benefit high-skilled, while disadvantaging low-skilled workers. The impact of such models on labour relations and on the working conditions faced by rural workers in the semi-periphery, therefore, remains unaddressed, as does the fact that the countries in question are often important exporters of biomass.

Due to the potential of palm oil to gain strategic importance within the region, a discussion of the possible impact of the transition towards a bioeconomy on labour relations must include a closer look at the working conditions faced by migrant workers in the Malaysian palm oil sector. A focus on the dynamics and labour processes in emerging bio-based industries provides an opportunity to shed light on old and new patterns of social inequalities.

This chapter is structured as follows: (1) I start by explaining why and how I examine the exploitation of labour. (2) I then sketch out the core characteristics of the prevailing labour migration regime in Malaysia, (3) and follow with a presentation of my findings from the fieldwork I carried out between 2017 and 2019 in the east Malaysian state of Sabah, where I examined the working conditions of low-skilled migrant workers in the palm oil sector. (4) I discuss the results before concluding that the superexploitation of migrant workers constitutes an essential feature of migrant labour in the palm oil sector.

The empirical findings discussed in this chapter encompass expert interviews, 15 guided interviews with migrant workers employed by large medium-sized, and small plantations or processing companies (palm oil mills) and 5 guided interviews with oil palm smallholders, as well as participatory observations.[1]

10.2 Analysing Social Inequalities as Class Relations

From a politico-economic perspective, under capitalism, class antagonism is determined by the structural position that groups hold in relation to the means of production (Marx 1987 [1894]). Following Marx, two large classes exist in ideal terms—the working class and the capitalist class. Workers must sell their labour power in order to survive. The capitalist class possesses and controls the means of production and buys the labour power of the working class. In order to create surplus value, workers have to work more than they are paid in wages; this is referred to as surplus work (ibid., p. 231). As such, wages represent the value of the workers' labour power, not the value of what they produce. From the perspective of capital, wages are paid to workers to reproduce their labour power (ibid., p. 184). The surplus value generated through surplus work in the form of goods is extracted and re-invested in the cycle of accumulation by capitalists, which defines the fundamental logic of what Marx calls *exploitation* (ibid., p. 328).

Neo-Marxist approaches, which are advocated, among others, by Marxist-feminists and scholars concerned with the intersection between race/ethnicity and class, have added that not only wage labour, but also intertwined forms of informal and non-wage labour can be exploited to raise profits and/or keep reproduction costs at a minimum in capitalism (Dörre and Haubner 2012; Federici 2015; Wallerstein 1990). Non-wage labour can encompass unpaid care or subsistence work. On this basis, I argue that the relationship between capital and various forms of labour is co-structured by the intertwined mechanism of superexploitation, which itself is based on social devaluation.

From the perspective of production, a hierarchical differentiation exists between wage and non-wage labour (Haubner 2017). This can

---

[1] I would like to thank Ramlah Binti Daud and Ryan Mukit for their support during fieldwork.

lead to a hierarchization of occupational groups within and outside the production sphere, which devaluates the labour power of a certain group compared to another (Roediger 2007). In order to give this devaluation social meaning, it is linked to and legitimized by (pre-existing) perceptions of, for example, gender or race/ethnic differentiations based on perceived or actual biological characteristics (Miles 1991, pp. 100–101) as well as status-related features such as citizenship (Wright 2015, p. 7). In the interests of capital, the devaluation of a group can function as a mechanism that pushes down the price of this groups' labour power to below average (Marini 1974) or that appropriates work entirely without compensation. This is called *superexploitation*.[2]

Marini argues that superexploitation is a regular feature of wage labour in semi-peripheral countries sustaining capital accumulation in the industrialized centre (ibid.). I define superexploitation as the exploitation and appropriation of labour by capital that exceeds the extent of formally regulated exploitation within the sphere of wage labour. Formal wage labour follows the logic of the exchange of equivalents (Marx 1987 [1894], pp. 80–83)—meaning labour is exchanged for wages at a value that sustains the reproduction of labour power. This exchange is contractually governed and regulated by law. Superexploitation occurs when salaries earned through wage labour are not sufficient to reproduce the labour power of the workers who receive them (Delgado-Wise and Veltmeyer 2016, pp. 57–60, p. 86) and their dependents. In the case of labour migration, citizenship functions as a mechanism of social devaluation that enables superexploitation.

Whereas Marxist-feminists already view the appropriation of unpaid reproductive work that sustains the labour power of wageworkers as an integral part of all capitalist economies (Haug 2015); with superexploitation, I shift the focus slightly and concentrate on the exploitation and appropriation of different forms of labour within the production sphere. Before investigating the superexploitation of migrant workers in the Malaysian palm oil sector, it is particularly important to understand

---

[2] Examining the superexploitation of labour implies taking a closer look at the extraction of extra surplus value by capital. In this paper, I focus on empirically observable forms of superexploitation and leave aside a critical discussion of approaches that attempt to calculate the rate of superexploitation or extra surplus value.

two issues: first, the institutional framework determining the position of migrant workers on the Malaysian labour market, which is based on citizenship, and, second, how this position translates into specific working conditions for migrant workers within labour processes.

10.3 Migratory Work in Malaysia: The State's Labour Migration Regime

Already under colonial rule, foreign labour became an integral part of Malaysia's economic development. Local labour was either unavailable or the native population refused to work under the harsh conditions of colonial capitalism (Garcés-Mascareñas 2012, p. 52). The demand for external sources of labour intensified at the beginning of the twentieth century with the steady growth of key economic sectors such as the production of natural rubber (ibid., p. 6), which was largely replaced by oil palm in the 1960s (Pichler 2014, p. 92). Today, Malaysia has the fourth largest migrant worker population in the world (Kotecha 2018, p. 2) and counts as the biggest 'net-importer' of foreign labour in the region (Ford 2014, p. 311). Approximately a quarter of the total workforce consists of migrants from the region (ILO—International Labour Organization 2016). Migrant workers are primarily hired to perform 'low-skilled' jobs (Sugiyarto 2015, p. 281). Political efforts to bridge the persistent gap between the supply and demand for cheap labour in Malaysia and state support for out-migration provided by sending countries as well as the choice by foreign workers to seek employment in Malaysia have led to a gradual formation of a transnational reserve of migrant labour power (Ferguson and McNally 2015, p. 3).

Although the role of foreign labour must be analysed in its historic-specific context if the political economy of labour migration in Malaysia is to be understood, certain characteristics have solidified over time: since the 1970s, the influx of labour migrants has led to the formation of a state-regulated labour migration regime with a highly segmented labour market (Ford 2014; Garcés-Mascareñas 2012, p. 56). Political measures promoting this segmentation include the channelling of low-skilled migrant workers into what are viewed as dirty, dangerous and degrading jobs, discrimination on the labour market, while, at the same time, providing support for skill development, further training and higher education to Malaysians. Capitalists have continuously wielded their power to ensure the migrant labour supply remains flexible and to keep their wages low. Today, the relative proportion of low-skilled migrant workers in the labour market depends on the cyclical economic demand for cheap labour and the political influence of nationalist, employee-friendly actors to limit labour migration in favour of the domestic workforce (ibid., p. 196). To understand how the broad institutional framework of the Malaysian labour migration regime is constructed in the workplace, it is necessary to take a closer look at the regulation of migrant labour.

Malaysia prevents migrant workers from establishing a life beyond their work, unless they have been granted a Malaysian identity card, which is extremely rare (SPIEU, Interview no. 5). Migrant workers are not allowed to bring their families with them or marry in Malaysia (Pye et al. 2012, p. 331), but they often either ignore this restriction by bringing their family members with them illegally or by bypassing the law by faking the birth certificates of their children or bribing state officials to grant family members access to the country.

Migrants who seek work in Malaysia must apply for a formal working permit, which is initially valid for three years and can be extended by up to two years. The state grants different types of permits to nationals from certain countries to work in selected branches of the economy, which results in the state-regulated division of labour by citizenship (Khoo 2001, p. 181). As migrant workers are not allowed to change jobs once a permit has been granted, they become highly dependent on their employer (Pye et al. 2016). If workers switch jobs without permission, when their working permits are withdrawn because of an economic recession or when they expire and workers choose to re-/enter or stay in Malaysia without valid documents they are drawn into illegality (ibid.). Their position on the labour market then changes in two ways: on the one hand, undocumented workers gain autonomy, as they are now free to 'move from one job to another, they do not pay taxes and it is […] difficult to make them leave the country' (Garcés-Mascareñas 2012, p. 84).

On the other hand, they risk being caught by state authorities or vigilante groups and sent to detention centres, where they may experience corporal punishment or food shortages, and will eventually be deported (Pye et al. 2012, p. 332). In the palm oil sector, many employers seize workers' passports to prevent them from running away or claim to do so as part of security measures. However, it is crucial that migrant workers hold on to their own passport as it is essential for freedom of movement.

The Malaysian state externalizes its reproduction costs to the countries from which it receives foreign workers (Pye 2014, p. 193), as well as to private companies and non-profit organizations. For example, as migrant children are not allowed to attend state schools in Malaysia, they must go to a school sponsored by a non-profit organization or be sent back to their country of origin in order to attend school, where they either remain on their own or female family members take care of them.

The following section shows how the Malaysian labour migration regime translates into the superexploitation of migrant workers within the production sphere of the palm oil sector by contextualizing my own qualitative investigation with empirical findings from other researchers.

10.4 Working Conditions of Migrant Plantation and Mill Workers

The palm oil sector heavily relies on the cheap labour of migrant workers in order to keep palm oil profitable and globally competitive (Pye et al. 2012, p. 331). In 2012, 73% of all workers employed in the palm oil sector worked as harvesters, loose fruit collectors or field workers on oil palm plantations. Around 87% of these workers were non-Malaysian (ibid.), and most of them were from Indonesia (Pye 2013, p. 10). Workers migrating to Malaysia are primarily 'attracted by […] higher wages' and the 'hope to save enough money […] to improve their livelihood possibilities at home' (ibid.). However, migration can be costly (Lindquist 2017) and salaries hardly ever exceed the minimum wage (Ford 2014; ILO 2016). This also applies to mill workers, who usually have a migrant background as well.

Studies have shown that great variations exist in the wage systems that are applied within the palm oil sector. These range from permanent contract-based salaries, wages based on harvesting quotas or piece rates to daily wages (Pye et al. 2016). In order to understand variations in wages and working conditions, it is important to draw a distinction between different types of employers.

10.4.1 Un(der)Paid, Underemployed and Undocumented

In 2016, around 61% of oil palm plantations were operated by private estates; independent smallholders made up for little more than 16% and government and state schemes planted around 22.5% of the total oil palm area (MPOB—Malaysian Palm Oil Board 2016). While large private estates and mills shape the agro-industrial mode of production in the sector (Cramb and McCarthy 2016, p. 53), smallholders have very little influence on the production model and are less resilient when faced with rapid global market developments. The production process on large estates follows a strict and highly gendered division of labour (Pye et al. 2016). While male migrant workers carry out physically demanding tasks and operate heavy machinery, female workers mostly spray fertilizer, collect loose fruit or work in the oil palm nursery. As female workers are mainly hired as daily workers at the lowest rate of pay in the industry, they are the most vulnerable workgroup (ibid.) to superexploitation.

Bigger and medium-sized companies usually provide workers with basic training, safety briefings and protective gear. By contrast, migrant workers employed by smallholders regularly perform tasks autonomously without guidance or monitoring from plantation owners. In many cases, they do not receive safety equipment or training, and are forced to rely on their experience or self-taught skills. Workers who work for smallholders perform multiple tasks, some of which are unpaid, and working hours remain undocumented. One respondent even mentioned that when he first arrived in Malaysia, he initially worked for a smallholder for free in order to have a place to stay and gain a foothold in Malaysia (male plantation worker, Interview no. 1). Other respondents explained that there was not always enough work for them because of the limited plantation size or seasonal factors.

The income situation of workers employed by smallholders can be particularly precarious because their basic salary varies depending on the employers' willingness and ability to pay, as the following statement exemplifies:

> My future depends on how many fruit bunches I can harvest. […] To me, the pay is not enough. […] But I also think about my employer. He […] can't afford to pay me the minimum wage […] I've spoken with my employer a few times about increasing my salary but it is still the same. (ibid.)

During the fieldwork, I found that the wages paid by smallholders were always below the minimum wage. Workers are often paid by piece rate, which means that their salaries depend on their productivity. To achieve fixed harvesting targets, migrant workers commonly involve family members—including minors (TFT, Interview no. 6). This additional labour power is not paid, yet it remains essential as workers cannot always cope with the workload.

In larger companies, salaries are paid on a contractual basis, or, in the case of organized labour, they are regulated by a collective agreement. Even though employers officially pay a minimum wage of 920 MYR (approx. 230 USD),[3] fieldwork revealed that deductions (e.g. levies for passports) meant that actual wages for migrant workers were lower. Formally at least, larger companies have fixed working hours and rules concerning overtime. Nevertheless, migrant workers either relied on overtime to increase their monthly income to sustain or enhance their family's livelihood or they were forced to work longer hours when production temporarily increased. Thus, overtime is more of a norm rather than an exception.

---

[3] In comparison, in 2016 the average monthly household income in Malaysia was 5.228 MYR (approx. 1,238 USD). See, https://www.dosm.gov.my/v1/index.php?r=column/ctwoByCat&parent_id=119&menu_id=amVoWU54UTl0a21NWmdhMjFMMWcyZz09. Accessed 22 April 2020.

Undocumented workers in the sector are especially vulnerable to wage-dumping (Pye et al. 2016). They depend on employers who are willing to provide them with work and on their relatives and friends to help them to stay under the radar. Workers who decide to or who are forced to work without a work permit may gain autonomy by working for the employer who pays them the highest wages, as the following statement by a worker verifies:

> I am an Indonesian citizen, if an employer pays one Ringgit[4] but another employer pays two, of course I will go. I want to earn more. (male plantation worker, Interview no. 2)

However, without legal regulation, wages are a matter of negotiation between the employer and the worker, and undocumented workers have very little bargaining power. Furthermore, undocumented migrant workers become highly dependent on a social network that (financially) supports them. They may even depend on the employment of legal migrant workers as illustrated by one case, in which a female plantation worker helps her friend who works for a subcontractor by sharing her salary:

> I was supposed to leave Malaysia in March but I do not want to leave […] I probably won't be here for long [anymore]. But my friends are helping me, even though it is illegal. (female plantation worker, Interview no. 3)

The working conditions and financial situation of migrant workers employed by different types of employers imply that workers must develop coping strategies to maintain the long-term reproduction of their labour power. To understand these strategies in relation to superexploitation, the next section deals with the connection between income precarity, the struggle to reproduce workers' labour power and the households they are part of, as well as various forms of labour-enabling reproduction.

---

[4]Ringgit is the national currency of Malaysia.

10.4.2 Struggling to Reproduce Livelihoods

I refer to the *household* as the 'unit that pools income for purposes of reproduction' (Wallerstein and Smith 1992, p. 15). As I have argued elsewhere, in order to understand the socio-economic situation of migrant workers in the palm oil sector, the *two-level family household* (Puder 2019, pp. 39–40) must be taken into account. I distinguish between the *nuclear family household*, consisting of family members forming a household unit in Malaysia, and the *transnational extended family household*, which includes relatives within and outside Malaysia (ibid.; Wallerstein and Smith 1992, p. 4).

Empirical findings have shown that income from wage work in the palm oil sector is distributed within the nuclear family household to secure its immediate reproduction. Due to the lack of employment opportunities for rural workers in their country of origin (Li 2009), migrant workers employed in the Malaysian palm oil sector might also contribute to the reproduction of the extended family by sending remittances to their family members. How the income is distributed among the two-level family household depends on the composition of the family as well as the relative proportion of family members who are able to contribute to the household income. The income situations of larger nuclear family households with fewer people in wage work are more precarious, and they are less able to send remittances regularly to their extended families or reproduce their labour power. An extreme example of this is illustrated by the case of a male mill worker with six children. He stated that his family regularly suffers from food shortages as his salary is just enough to buy basic foods such as rice, sugar and salt. He can only buy fresh fish and vegetables for his family if he does overtime or finds other sources of income (male mill worker, Interview no. 4). This case demonstrates that working overtime to increase the household income or to acquire savings is integral to the typical working week of migrant workers and that it blurs the lines between regular working hours and overtime. In contrast, smaller household units with more people contributing to total household income are more likely to be able to send remittances frequently to the extended family household. In this context, remittances can represent an essential feature of the reproduction of transnational families (Pye et al. 2012, p. 332).

Low wages mean that migrant workers who work for smaller companies or smallholders must find additional sources of income to provide for their basic needs. Almost all respondents stated that they had experienced income insecurity in the past and still do so. Therefore, migrant workers regularly shift between formal wage work and informal or subsistence work. To cope with income insecurity, female family members engage in activities such as selling pastries or fruit to staff or on the weekly market. Male workers employed by smallholders often take up additional, informal harvesting or maintenance jobs on other estates, which their main employers either tolerate or encourage by recommending their workers to fellow smallholders.

Migrant workers try to overcome their income precarity by asking friends and relatives to help them find a better-paid job or a family member to support them financially. They also attempt to establish a network to share information about the wages paid by different employers. If migrant workers' households run out of money because of unexpected costs (e.g. fixing a car), or rising living expenses, a common coping strategy is to either borrow money or buy groceries using a buy-now-pay-later system. Such systems may even be institutionalized by large companies: one respondent explained that if workers purchase food on credit from a small store on the estate, the company deducts the outstanding payment from the workers' next pay cheque (male plantation worker, Interview no. 2). This system can normalize securing a minimum standard of living through debt if precarious workers regularly rely on the buy-now-pay-later system to satisfy their basic needs.

The reproduction of wage labour in terms of care work also rests on the support system provided by the two-level family household. While daily care work in the nuclear family household is mostly performed by female family members, the transnational extended family household functions as a cross-border network of reproduction (Pye 2014, p. 195). In such cases, the relatives provide a substitute for the lack of access to social welfare in Malaysia and the restrictions on welfare services available in the country of origin. Consequently, the externalization of reproduction costs, as part of the Malaysian labour migration regime, leads to an appropriation, not only of female care work within the nuclear family household but also of non-waged care work enacted by members of the extended family in and outside of Malaysia.

In Malaysia, migrant workers perform subsistence agriculture if they have access to small areas of land that are either provided by their employers or that they occupy to secure their livelihood. In their country of origin, family members mostly use the remittances their relatives send from Malaysia to buy and cultivate land for subsistence farming.

Generally, the division of labour in the household can be understood as a strategy that strengthens their social security by helping its members to deal with precarious working conditions and low wages. At the same time, the practice of income distribution within the two-level family household as well as the engagement of its members in various forms of labour reproduces the Malaysian migrant labour regime in the palm oil sector—and, in doing so, the constant struggle to reproduce livelihoods. This is reinforced by the robust barriers to workers' struggle that are outlined in the following.

10.4.3 Barriers to Workers' Struggle

The strong hurdles faced by migrant workers and their weak bargaining power when organizing are further factors that contribute to the super-exploitation of migrant workers.[5] In many cases, migrants are unaware of their legal status and their basic rights. As workers are often unsure about the content of the documents they signed when starting to work in Malaysia or do not even know whether they signed a contract at all, they become especially vulnerable to employers who are able to abuse these knowledge gaps (SPIEU, Interview no. 5). Illiteracy, a lack of experience in enforcing their rights and a fear of losing their job means that some workers simply accept whatever conditions that employers offer.

---

[5] Pye argues that migrant workers' struggles emerge in the form of *everyday resistance* (2017). I solely concentrate on collective action on the macro- rather than on the micro-level.

Similarly, the majority of workers are unaware of their right to join a union or to demand safety gear. In addition, if workers are unable to carry out their tasks because of rainfall, sickness or because a family issue requires them to return to their country of origin, they do not get paid. The exception was one case where a collective bargaining agreement by a union led the workers to enjoy these kinds of 'privileges'.

It is extremely difficult for unions to initiate organizing because plantations and mills are generally located in remote areas, it is difficult to enter large estates without permission from the owner, and workers who work for smallholders are usually isolated from one another. Furthermore, illegal migrant workers usually have no union representation at all because it is difficult to represent them and they also worry that they will be deported if they demand better working and living conditions (Assalam 2019). Legal obstacles also act as barriers to workers' struggles and limit unionization. Even if a union is successful in organizing documented workers, unions must ensure that a 50% plus 1 (50 + 1) majority of *all* employees in an entire company have joined the union to gain official company recognition and to be able to enter into collective bargaining (SPIEU, Interview no. 5). In the past, large companies have found various ways of preventing their employees from organizing by denying unions access to mills and plantations, threatening to fire rebellious workers and intimidating organizers (ibid.; SPN, Interview no. 7).

10.5 Conclusion: Bioeconomy as a Continuation of Superexploitation?

This chapter discussed social inequality in labour relations in the wake of emerging bio-based industries as exemplified by the case of palm oil and its links to labour migration in Malaysia. I argued that the Malaysian palm oil sector rests on the superexploitation of migrant workers and that this is made possible by the social devaluation of this social group due to their (lack of Malaysian) citizenship. This devaluation leads migrant workers to face political, legal and socio-economic discrimination, which, in turn, makes them the most precarious group in the industry. The state regulation of labour migration keeps reproduction costs for the state low and establishes a framework in which companies can maximize the exploitation of migrant labour. Hence, the labour migration regime paves the way for capital owners in the palm oil industry to superexploit migrant workers.

Drawing upon the empirical findings, no matter which type of employer they work for, workers are sometimes paid wages that are below the minimum wage. While larger companies try to stretch legal regulations concerning minimum wages and overtime, workers employed by smallholders are regularly un(der)paid and underemployed. As a consequence, migrant workers are unable to reproduce their labour power or their two-level family households by solely relying on formal employment. The superexploitation of migrant workers manifests itself in salaries that are below a living wage, but also in the appropriation of informal, subsistence and reproductive work of both household levels, and the constant pressure to perform overtime and in recurring income insecurities. The externalization of the costs of reproduction, therefore, must be compensated for by the household's cross-border support network. As such, the household not only bears the costs of social reproduction but also the work that has to be done to carry it out. Undocumented workers are fully exposed to un(der)payment and underemployment irrespective of the type of employer they work for. Furthermore, the absence of legal institutions guaranteeing compliance with minimum working standards places undocumented migrants at risk of even worse forms of superexploitation.

To sum up: the superexploitation of migrant workers in the Malaysian palm oil sector—a possible key sector in a future global bioeconomy—demonstrates that we must take a closer, critical look at state policies that promote bio-based industries and to ensure that they not only promise the greening of the economy but also better working and living conditions for the workers employed in relevant sectors.

List of Interviews quoted

Interview no.	Gender and job position/Organization	Date and place
Interview no. 1	Male plantation worker	09/03/2018, Sandakan
Interview no. 2	Male plantation worker	14/03/2018, Tawau
Interview no. 3	Female plantation worker	14/03/2018, Tawau
Interview no. 4	Male mill worker	15/03/2018, Kunak
Interview no. 5	SPIEU	14/03/2018, Tawau
Interview no. 6	TFT	10/04/2018, Kuala Lumpur
Interview no. 7	SPN	10/05/2019, Phone interview

References

Anderson, Z.R. (2016). Assembling the 'Field': Conducting Research in Indonesia's Emerging Green Economy. *Austrian Journal of South-East Asian Studies, 9*(1), 173–180.

Assalam, R. (2019). Herausforderungen bei der Organisierung von Arbeitsmigrant*innen in den Palmölplantagen Sabahs. https://suedostasien.net/herausforderungen-bei-der-organisierung-von-arbeitsmigrantinnen-in-den-palmoelplantagen-sabahs/. Accessed 13 April 2019.

Backhouse, M., Lorenzen, K., Lühmann, M., Puder, J., Rodríguez, F., & Tittor, A. (2017). Bioökonomie-Strategien im Vergleich: Gemeinsamkeiten, Widersprüche und Leerstellen. *Working Paper 1, Bioeconomy & Inequalities*, Jena. https://www.bioinequalities.uni-jena.de/sozbemedia/neu/2017-09-28+workingpaper+1.pdf. Accessed 18 March 2020.

Birch, K. (2019). *Neoliberal Bio-economies? The Co-construction of Markets and Natures*. Cham: Palgrave Macmillan.

Brand, U., & Wissen, M. (2015). Strategies of a Green Economy, Contours of a Green Capitalism. In K. van der Pijl (Ed.), *Handbook of the International Political Economy of Production* (pp. 508–523). Cheltenham: Edward Elgar Publishing.

Choong, Y.Y., Chou, K.W., & Norli, I. (2018). Strategies for Improving Biogas Production of Palm Oil Mill Effluent (POME) Anaerobic Digestion: A Critical Review. *Renewable and Sustainable Energy Reviews, 82,* 2993–3006.

Cramb, R., & McCarthy, J.F. (2016). Characterising Oil Palm Production in Indonesia and Malaysia. In R. Cramb & J.F. McCarthy (Eds.), *The Oil Palm Complex: Smallholders, Agribusiness and the State in Indonesia and Malaysia* (pp. 27–77). Singapore: NUS Press.

Delgado-Wise, R., & Veltmeyer, H. (2016). *Agrarian Change, Migration and Development.* Warwickshire: Fernwood Publishing.

Dörre, K., & Haubner, T. (2012). Landnahme durch Bewährungsproben: Ein Konzept für die Arbeitssoziologie. In K. Dörre, D. Sauer, & V. Wittke (Eds.), *Kapitalismustheorie und Arbeit: Neue Ansätze soziologischer Kritik* (pp. 63–106). Frankfurt a.M.: Campus.

Federici, S. (2015). *Caliban and the Witch: Women, the Body and Primitive Accumulation.* New York: Autonomedia.

Ferguson, S., & McNally, D. (2015). Precarious Migrants: Gender, Race and the Social Reproduction of the Global Working Class. *Socialist Register,* 1–23.

Ford, M. (2014). Contested Borders, Contested Boundaries: The Politics of Labour Migration in Southeast Asia. In R. Robison (Ed.), *Routledge Handbook of Southeast Asian Politics* (pp. 305–314). New York: Routledge.

Garcés-Mascareñas, B. (2012). *Labour Migration in Malaysia and Spain: Markets, Citizenship and Rights.* Amsterdam: Amsterdam University Press.

Goven, J., & Pavone, V. (2015). The Bioeconomy as Political Project: A Polanyian Analysis. *Science, Technology, & Human Values, 40*(3), 302–337.

Haubner, T. (2017). *Die Ausbeutung der sorgenden Gemeinschaft: Laienpflege in Deutschland.* Frankfurt a.M.: Campus.

Haug, F. (2015). *Der im Gehen erkundete Weg: Marxismus-Feminismus.* Berlin: Argument.

ILO (2016). Review of Labour Migration Policy in Malaysia. Tripartite Action to Enhance the Contribution of Labour Migration to Growth and Development in ASEAN (TRIANGLE II Project). Bangkok. https://www.ilo.org/asia/publications/WCMS_447687/lang--en/index.htm. Accessed 16 April 2020.

Ismail, A. (2013). The Effects of Labour Shortage in the Supply and Demand of Palm Oil in Malaysia. *Oil Palm Industry Economic Journal, 13*(2), 15–26.

Khoo, B.T. (2001). The State and the Market in Malaysian Political Economy. In G. Rodan, K. Hewison, & R. Robison (Eds.), *The Political Economy of South-East Asia. Conflicts, Crises, and Changes* (pp. 178–205). Oxford: Oxford University Press.

Kotecha, A. (2018). Malaysia's Palm Oil Industry. USAID & Winrock International. https://static1.squarespace.com/static/5592c689e4b0978d3a48f7a2/t/5b9a15db88251b25f1bc59d1/1536824861396/Malaysia_Analysis_120218_FINAL.pdf. Accessed 18 March 2020.

Li, T. (2009). To Make Live or Let Die? Rural Dispossession the Protection of Surplus Populations. *Antipode, 41*(1), 66–93.

Lindquist, J. (2017). Brokers, Channels, Infrastructure: Moving Migrant Labor in the Indonesian-Malaysian Oil Palm Complex. *Mobilities, 12*(2), 1–14.

Marini, R.M. (1974). Dialektik der Abhängigkeit. In D. Senghaas (Ed.), *Peripherer Kapitalismus: Analysen über Abhängigkeit und Unterentwicklung* (pp. 98–136). Berlin: Suhrkamp.

Marx, K. (1987 [1894]). *Das Kapital: Dritter Band* (MEW). Berlin: Dietz Verlag.

Miles, R. (1991). *Rassismus: Einführung in die Geschichte und Theorie eines Begriffs*. Hamburg: Argument.

MPOB (2016). Oil Palm Estates, January–December 2016. http://bepi.mpob.gov.my/images/area/2016/Area_summary.pdf. Accessed 18 March 2020.

OECD (2017). Employment Implications of Green Growth: Linking Jobs, Growth, and Green Policies: OECD Report for the G7 Environment Ministers. https://www.oecd.org/environment/Employment-Implications-of-Green-Growth-OECD-Report-G7-Environment-Ministers.pdf. Accessed 18 March 2020.

OECD (2019). The Bioeconomy to 2030: Designing a Policy Agenda. https://www.oecd-ilibrary.org/economics/the-bioeconomy-to-2030_9789264056886-en. Accessed 18 March 2020.

Obidzinski, K., Adriani, R., Komarudin, H., & Adrianto, A. (2012). Environmental and Social Impacts of Oil Palm Plantations and Their Implications for Biofuel Production in Indonesia. *Ecology and Society, 17*(1), 481–499.

Pichler, M. (2014). *Umkämpfte Natur: Eine politökologische Analyse der Rolle von Staatlichkeit in der Palmöl- und Agrartreibstoffproduktion in Südostasien*. Münster: Westfälisches Dampfboot.

Pichler, M. (2015). Legal Dispossession: State Strategies and Selectivities in the Expansion of Indonesian Palm Oil and Agrofuel Production. *Development and Change, 46*(3), 508–533.

Puder, J. (2019). Excluding Migrant Labor from the Malaysian Bioeconomy: Working and Living Conditions of Migrant Workers in the Palm Oil Sector in Sabah. *Austrian Journal of South-East Asian Studies, 12*(1), 31–48.

Pye, O. (2013). Introduction. In J. Bhattacharya & O. Pye (Eds.), *The Palm Oil Controversy in Southeast Asia: A Transnational Perspective* (pp. 1–18). Singapore: ISEAS Publishing.

Pye, O. (2014). Transnational Space and Worker's Struggles: Reshaping the Palm Oil Industry in Malaysia. In K. Dietz, B. Engels, O. Pye, & A. Brunnengräber (Eds.), *The Political Ecology of Agrofuels* (pp. 186–201). London: Routledge.

Pye, O. (2017). A Plantation Precariate: Fragmentation and Organizing Potential in the Palm Oil Global Production Network. *Development and Change*, 48(5), 942–964.

Pye, O., Daud, R., Manurung, K., & Siagan, S. (2012). Precarious Lives: Transnational Biographies of Migrant Oil Palm Workers. *Asia Pacific Viewpoint*, 53, 330–342.

Pye, O., Daud, R., Manurung, K., & Siagan, S. (2016). Workers in the Palm Oil Industry: Exploitation, Resistance and Transnational Solidarity. Köln. https://www.asienhaus.de/archiv/user_upload/Palm_Oil_Workers_-_Exploitation__Resistance_and_Transnational_Solidarity.pdf. Accessed 18 March 2020.

Roediger, D.R. (2007). *The Wages of Whiteness: Race and the Making of the American Working Class*. London: Verso.

Scarlat, N., Dallemand, J.-F., Monforti-Ferrario, F., & Nita, V. (2015). The Role of Biomass and Bioenergy in a Future Bioeconomy: Policies and Facts. *Environmental Development*, 15, 3–34.

Statista (2019). Palm Oil Industry in Malaysia. https://www.statista.com/study/63749/palm-oil-industry-in-malaysia/. Accessed 18 March 2020.

Sugiyarto, G. (2015). Internal and International Migration in Southeast Asia. In I. Coxhead (Ed.), *Routledge Handbook of Southeast Asian Economics* (pp. 270–299). New York: Routledge.

UBA (2014). Arbeit und Qualifikation in der Green Economy. https://www.umweltbundesamt.de/publikationen/arbeit-qualifikation-in-der-green-economy. Accessed 18 March 2020.

Varkkey, H. (2013). Malaysian Investors in the Indonesian Oil Palm Plantation Sector: Home State Facilitation and Transboundary Haze. *Asia Pacific Business Review*, 19(3), 381–401.

Wakker, E. (2005). Greasy Palms: The Social and Ecological Impacts of Large-Scale Oil Palm Plantation Development in Southeast Asia. London: Friends of the Earth. https://socialvalueuk.org/wp-content/uploads/2016/04/greasy_palms_impacts.pdf. Accessed 18 March 2020.

Wallerstein, I. (1990). Die Konstruktion von Völkern: Rassismus, Nationalismus, Ethnizität. In E. Balibar & I. Wallerstein (Eds.), *Rasse, Klasse, Nation: Ambivalente Identitäten* (pp. 87–106). Hamburg: Argument.

Wallerstein, I., & Smith, J. (1992). Households as an Institution of the World-Economy. In I. Wallerstein & J. Smith (Eds.), *Creating and Transforming Households: The Constraints of the World-Economy* (pp. 3–23). Cambridge: Cambridge University Press.

World Bank (2015). Malaysia Economic Monitor. Immigrant Labour December 2015. http://documents.worldbank.org/curated/en/753511468197095162/pdf/102131-WP-P158456-Box394822B-PUBLIC-final-for-printing.pdf. Accessed 18 March 2020.

Wright, E.O. (2015). *Understanding Class.* London, New York: Verso.

Open Access This chapter is licensed under the terms of the Creative Commons Attribution 4.0 International License (http://creativecommons.org/licenses/by/4.0/), which permits use, sharing, adaptation, distribution and reproduction in any medium or format, as long as you give appropriate credit to the original author(s) and the source, provide a link to the Creative Commons license and indicate if changes were made.

The images or other third party material in this chapter are included in the chapter's Creative Commons license, unless indicated otherwise in a credit line to the material. If material is not included in the chapter's Creative Commons license and your intended use is not permitted by statutory regulation or exceeds the permitted use, you will need to obtain permission directly from the copyright holder.

11

Sugarcane Industry Expansion and Changing Rural Labour Regimes in Mato Grosso do Sul (2000–2016)

Kristina Lorenzen

11.1 The Interrelations of Bioeconomy, Brazilian Sugarcane and Social Inequalities

The bioeconomy represents a future vision (Goven and Pavone 2015, p. 5), the actual realisation and societal consequences of which are difficult to predict. This research centres around bioethanol. The production and use of bioethanol precede the bioeconomy discourse and, therefore, have already been fully implemented. Agrofuels such as bioethanol are being integrated into bioeconomy agendas (Backhouse et al. 2017, pp. 23–26).

Brazil is the second most important producer of bioethanol globally and has a tradition of commercial-scale sugarcane-based bioethanol production dating back to the 1970s (Wilkinson and Herrera 2010, pp. 750–757). The promotion of agrofuels is an important part of the

---

K. Lorenzen (✉)
Institute of Sociology, Friedrich Schiller University, Jena, Germany
e-mail: kristina.lorenzen@uni-jena.de

© The Author(s) 2021
M. Backhouse et al. (eds.), *Bioeconomy and Global Inequalities*,
https://doi.org/10.1007/978-3-030-68944-5_11

Brazilian bioeconomy.[1] The sugarcane sector started to expand around 2002 and was closely linked to the promotion of bioethanol production. This expansion dates back to a time in which bioeconomy was not discussed in Brazil. Nevertheless, it demonstrates what happens when the emergence and increased expansion of bioeconomy in Brazil include bioethanol as an important pillar.

Previous research has shown that the monocultural expansion of sugarcane can result in peasant displacement and conflicts over land, but it can also create employment opportunities (Borges et al. 1983, pp. 90–104). However, in this case, the expansion, which started around 2002, was accompanied by mechanisation of this area of the agricultural part of the sector, and, in the state of São Paulo, this led to reductions in the workforce (Brunner 2017, pp. 7–8). Hence, current research suggests looking at the interconnection of land *and* labour relations in order to understand the combined impact of geographical expansion and mechanisation. Against this background, the question arises as to how the expanding production of bioethanol—as an important pillar of the emerging (Brazilian) bioeconomy—affects existing social inequalities in labour and land relations.

This chapter applies a case study approach (Yin 2009). The selected case is sugarcane expansion between 2000 and 2016 in central southern parts of the Brazilian federal state of Mato Grosso do Sul, which demonstrates one of the most intensive dynamics of expansion in Brazil. The field research took place in April 2017, November and December 2017 and between April and June 2018. Open and semi-structured interviews, informal conversations and participatory observations were used, and a wide variety of people from civil society, academia, and the state and private sector were interviewed (see Lorenzen 2019).

The next section details the analytical framework. Section 11.3 describes the historical and recent dynamics of the Brazilian sugarcane sector and complements this with relevant global dynamics centring around green and sustainable development. Section 11.4 traces the

---

[1] Brazil has not yet developed a bioeconomy strategy. The Ministry of Science, Technology, Innovation and Communication addresses the bioeconomy in its "National Strategic Plan 2016-2022" (MCTIC—Ministério da Ciência, Tecnolgia, Inovação e Comunicações 2016, pp. 94–97).

changes in rural wage and subsistence work that has accompanied sugarcane expansion in the federal state of Mato Grosso do Sul. To close, Sect. 11.5 brings all the information together and assesses the changes in social inequalities by looking at rural labour regimes. Finally, some conclusions are drawn for bioeconomy policies.

11.2 Towards an Analytical Framework of Unequal Access to Labour and Land

11.2.1 Social Inequalities as Asymmetrical Access to Labour and Land

Focusing on the unequal access to land and labour of different social groups, I understand social inequalities to refer to the systematic asymmetrical and hierarchical access of groups to aspects such as economic goods (work, income) and natural resources (land) that leads to beneficial or disadvantageous living conditions (Solga et al. 2009, p. 15; Kreckel 2004, p. 17). In line with the Theory of Access, access is understood as "the ability to benefit from things" (Ribot and Peluso 2003, p. 153).

This study focuses mainly on the national state of Brazil and the local context. As such, it takes the federal level of Brazil and the state of Mato Grosso do Sul as its point of departure. The analysis unfolds from a historically informed perspective and moves to study how the expansion of ethanol affects social inequalities in terms of land and labour. This presupposes an account of the transnational context in which these relationships are embedded.

11.2.2 Labour Regimes as Combining Access to Labour and Land

To conceptualise how land and labour relations interconnect, I draw on agrarian political economy and especially on its conception of rural labour regimes. Within this strand of thought, Henry Bernstein (2010, pp. 21–26) points out that the most distinctive feature of capitalism is the social relation of exploitation in which capitalists own the means of production and labourers have to sell their labour power to obtain their subsistence. This basic understanding of capitalism has sometimes led to the idea that all peasants will eventually be dispossessed and proletarianised in capitalism. This idea underestimates the complexities of social relations of global capitalism. However, Bernstein argues further that "capital is capable of exploiting labour through a wide range of social arrangements in different historical circumstances" and that categories such as "landless labour" and "small peasants" are fluid and people move between those categories or can be categorised as both at the same time. Proletarians are not the only group to be exploited by capital and capitalism does not necessarily always result in the dispossession of small peasants (ibid., pp. 33–34).

How then can we understand non-wage labour categories as part of capitalism? Proponents of the subsistence approach (von Werlhof et al. 1988) coined the term subsistence production for all work that is not wage labour. Subsistence production is understood as both the production of life and the production of means of subsistence. Thus defined, subsistence work also includes the activities of peasants (Mies 1988, p. 86). Although these activities have often been called pre-capitalist, materialist feminism as well as dependency theories and world system analysis have shown that non-wage work such as housework and small-scale agriculture have been produced or reconfigured by capitalism as a means of outsourcing the costs of social reproduction (Boatcă 2016, pp. 74–75; Bohrer 2018, p. 65; Federici 2012, p. 22; Mies 1988). In the field of subsistence production, my research focuses on the production of the means of subsistence, especially through non-wage agriculture, hunting and fishing.

The combination of access to different categories of rural labour and the varying combination of access to wage labour and access to land for subsistence production results in different labour regimes. The concept of the labour regime refers to "different methods of recruiting labour and their connections with how labour is organised in production (labour processes) and how it secures its subsistence" (Bernstein 2010, p. 53). While some analyses of labour regimes focus on the methods of recruiting labour (coercive/non-coercive) or on the details of how labour is organised in production (see e.g. Brunner 2017; Li 2017), I focus on how different categories of labour are combined to secure subsistence. "Pure" labour regimes exist that only provide access to wage labour (proletarianisation) *or* access to land (peasant production). And there are "hybrid" forms such as semi-proletarianisation (Bernstein 2010, pp. 53–55). Semi-proletarianisation describes a mixed labour regime, where land access for subsistence activities is combined with access to wage work to secure livelihoods (Boatcă 2016, pp. 65–66). This approach of labour regimes to agrarian political economy serves to identify changes in the interdependency of land and labour relations. However, before these changes are assessed, the next chapter describes the recent changes that have occurred in the Brazilian sugarcane sector.

11.3 The Brazilian Sugarcane Sector and Its Recent Changes

Sugar was one of the first products exported by Brazil when it was a Portuguese colony (Baer 2014, p. 14). The use of sugarcane to produce ethanol for bioenergy is more recent. On a commercial scale, ethanol production started in 1975 with the implementation of the National Ethanol Programme *Proálcool*. As part of the Proálcool programme, the government subsidised the establishment of eight industrial distilleries in Mato Grosso do Sul (Missio and Vieira 2015, pp. 179–180; Domingues 2017, p. 76). However, Mato Grosso do Sul was yet to become an important bioethanol producer.

This started to change at the beginning of the new millennium when the dynamics of global land grabbing fostered investment in land. In Brazil, large-scale land acquisitions increased after 2002. The sugarcane sector was one of the most important sectors for these land deals (Borras et al. 2011, p. 17; Sauer and Leite 2011, p. 1). At the same time, national policies such as a blending quota, subsidised loans and tax benefits facilitated the resurgence of the crisis-ridden sugarcane sector in Brazil (Sant'Anna et al. 2016a, pp. 166–167; Wilkinson 2015, p. 3).

State support for the ethanol sector can be explained from the perspective of a global bioethanol market as this was driven by demands that rose out of the Kyoto Protocol and COP21 (Wilkinson 2015, pp. 2–3). In 2007, the Brazilian government published a study indicating that Brazil could supply 5% of the world's consumption of car fuel (Defante et al. 2018, p. 126). Nevertheless, for Brazilian bioethanol to become "green", it had to undergo change. The Brazilian government had to ensure that sugarcane production was "sustainable" enough to be viewed as an alternative energy-source for transportation worldwide. Therefore, it promoted zoning-projects, which led to the exclusion of sensitive and biodiversity-rich areas from land investments. Furthermore, the government and the sugarcane industry agreed on a protocol (*Protocolo Agroambiental*) that abolished the practice of burning sugarcane before harvesting. Although burning facilitates the harvest, it also releases large amounts of CO_2. The solution, therefore, was to gradual mechanise the sugarcane harvest and to abandon the use of burning (Jesus and Torquato 2014; Wilkinson 2015, pp. 2–3).

The entanglement of global land grabbing dynamics and national policies led to a boom and an expansion of sugarcane in Brazil. The boom caused land prices to rise in the main area used for cultivation: the federal state of São Paulo. Investors left for neighbouring states such as Mato Grosso do Sul, where the number of production units rose from eight to 22 (Assunção et al. 2016, pp. 6–7).

In 2010 and 2011, when the global financial crises hit Brazil, credit programmes were cut and the unsustainably financed sugarcane sector partly collapsed. Production units closed or were bought by international investors, which concluded the process of internationalisation that had already begun (Wilkinson 2015, p. 3). In spite of the crisis, the area of land used for sugarcane and the production of sugarcane, sugar and ethanol continued to increase (see Observatório da Cana[2]; Lorenzen 2019, p. 19). Since the 2014/2015 harvest, the sugarcane sector has been slowly recovering.

The expansion and mechanisation of the Brazilian sugarcane sector were triggered by the interrelations of global and national dynamics. These dynamics had an important yet ambivalent impact on local land

---

[2] https://observatoriodacana.com.br/historico-de-area-ibge.php?idMn=33&tipoHistorico=5. Accessed 12 Nov 2020.

and labour relations in Mato Grosso do Sul, and these will be illustrated in the next section.

11.4 The Impact of the Expansion of the Sugarcane Industry on Access to Labour and Land

In the following, the changes in access to labour and land are described using the two social groups that were the most affected by the expansion and mechanisation of the sugarcane industry. These descriptions are brought together in Sect. 11.5 as part of the analysis of the associated labour regimes.

Peasants in agrarian reform settlements constitute the first group. There are two main types of peasants in Brazil: small-scale agriculturists who own property, and peasants who have obtained access to public land via the agrarian reform process (Damasceno et al. 2017, p. 18). The agrarian reform process in Brazil is aimed at expropriating private land that is no longer put to "productive"[3] use (Fernandes et al. 2010, p. 799). Expropriated land is then turned into agrarian reform settlements with smaller lots that are transferred to landless workers.[4]

The Guarani-Kaiowá Indigenous people constitute the second group. Mato Grosso do Sul is the Brazilian federal state with the second-largest Indigenous population: 3% of its population describes themselves as Indigenous people.[5] The Guarani-Kaiowá form the largest ethnic group

---

[3] I use the term unproductive land in accordance with the INCRA (Instituto Nacional de Colonização e Reforma Agrária) definition: "The INCRA considers property (rural property) to be unproductive in cases where arable land is either totally or partially unused by its occupant or owner" (Author's translation, INCRA. See, http://www.incra.gov.br/pt/educacao/2-uncategorised/233-imovel-rural-improdutivo.html. Accessed 12 Nov 2020.)

[4] INCRA, Obtenção de terras: see, http://www.incra.gov.br/pt/obtencao-de-terras.html. Accessed 13 Nov 2020. And Assentamentos: see, http://www.incra.gov.br/pt/assentamentos.html. Accessed 12 Nov 2020.

[5] FUNAI (Fundação Nacional do Índio): see, http://funai.gov.br/index.php/comunicacao/noticias/1069-entre-1991-e-2010-populacao-indigena-se-expandiu-de-34-5-para-80-5-dos-municipios-do-pais. Accessed 5 Sep 2018.

within this population.[6] The 1988 Brazilian Constitution recognises the right of Indigenous people to their traditional lands and obliges the government to demarcate these lands. Unfortunately, competing interests were already present by the time the government started demarcating Indigenous land. This resulted in uncertainty and conflict over land due to overlapping property, a situation that persists to this day (Damasceno et al. 2017, p. 18). The Guarani-Kaiowá in the central south of Mato Grosso do Sul mainly live in Indigenous reservations, on small demarcated Indigenous lands or on occupied land which is being reclaimed (*retomadas*). In the next section, I trace the changes in wage work in the sugarcane sector.

11.4.1 Wage Work: Mechanisation, Employment Creation and Unemployment

One major impact of the expansion of the sugarcane industry is the creation of employment prospects. Other rural industries in Mato Grosso do Sul such as cattle raising and soybean production are less labour intensive[7]; thus, the expansion of the sugarcane sector provided new job opportunities (Fig. 11.1). In 2016, the sugarcane sector in Mato Grosso do Sul employed 25,577 people in industry and agriculture, a figure that corresponded to 1.2% of all people in employable age (2,130,000 individuals, SEMAGRO 2018). After 2012, there was a drop in the number of people employed in this sector. This was partly due to the financial and economic crisis.

Figure 11.2 highlights another reason for the decline in the number of people employed in this sector. The figure shows the difference in the number of employees in agriculture (sugarcane cultivation and harvest) and industry (the production of sugar and ethanol). While the number of industrial workers increased until 2014 and then only declined slightly, the number of agricultural workers declined steadily. By 2012, most

---

[6] Museu das Culturas Dom Bosco: see, http://www.mcdb.org.br/materias.php?subcategoriaId=23. Accessed 5 Sept 2018.
[7] Big landowner who cultivates soybeans, Interview no. 1; university professor, Interview no. 2; person from an organic agriculture association, Interview no. 3.

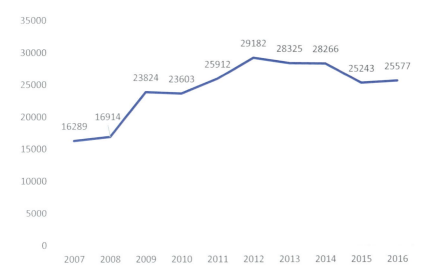

Fig. 11.1 Employees in the sugarcane sector in Mato Grosso do Sul, 2007–2016 (*Source* RAIS, organised by DIESSE)

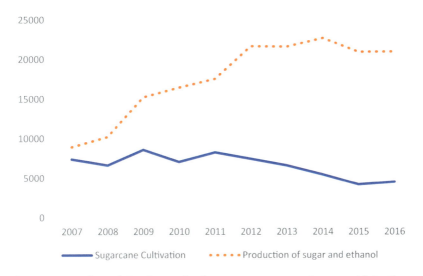

Fig. 11.2 Number of Employees in the sugarcane sector by area, Mato Grosso do Sul, 2007–2016 (Source RAIS, organised by DIESSE)

sugarcane harvesting and planting in Mato Grosso do Sul was mechanised (MPT, Interview no. 4). One harvesting machine is estimated to replace up to 100 workers (Abreu et al. 2009, p. 6). The Indigenous people who had worked mainly in agriculture were laid off or no longer hired.

During the implementation of the Proálcool programme, the Guarani-Kaiowá were the most important labour force and continued to be so until around 2012 (MPT, Interview no. 4). They also acted as a cheap source of labour, were not provided with formal contracts and the companies did not comply with basic labour rights. The working conditions were often slave-like as defined in the Brazilian criminal code, and the workers faced undignified conditions, excessive workloads, as well as forced and bonded labour (Pauletti 2014, pp. 53–59; Repórter Brasil 2017; de Rezende 2014, pp. 195–198). Working conditions started to improve, when a regional department of the Public Prosecution for Labour Rights (MPT) was founded. Fines were and continue to be imposed on companies that hold workers in such conditions (Pauletti 2014, pp. 39–42; de Rezende 2014, pp. 198–199).

However, profound improvements were only made to working conditions with the spread of mechanisation. The problems with the working conditions were primarily the inadequate board and lodging that hundreds of temporal Indigenous labourers faced. Recruitment and hiring policies changed with mechanisation. Instead of hiring hundreds of temporal labourers, which were housed on the edges of the sugarcane plantations for weeks or months, they permanently employed a smaller number of people who lived near the company as truck drivers and machine operators. This made board and lodging unnecessary. Additionally, the companies started to employ higher qualified personnel and were therefore willing to offer formal contracts, provide better wages and additional benefits.[8] The regional department of the Public Prosecution for Labour Rights (MPT) reports that the problems with slave-like labour in the sugarcane sector have declined tremendously (MPT, Interview no. 4).

---

[8]These benefits and the better payment have been under attack since the labour reform in 2017, which means that working conditions have deteriorated since then (representative of a labour union, interview no. 5).

However, Indigenous people did not benefit from these improved working conditions since they mainly lost their access to wage labour in the sector. Instead, Indigenous people reported that they were never told why they were no longer being hired. Representatives of labour unions assumed that they were not qualified or did not have a sufficient level of education to work as a tractor or truck driver, machine operator or in the industrial sector (representative of a labour union, Interview no. 5; representative of a labour union, Interview no. 6). Furthermore, some Indigenous people suspected that as they had been involved in years of public prosecutions and attempts to enforce the law to achieve better working conditions, the companies were no longer willing to hire them.[9] Certainly, there is deeply rooted discrimination against Indigenous people in Brazil and one of the most common prejudices is the perception that they are lazy.[10]

While Indigenous people mostly lost their access to the sugarcane sector, new access to employment opportunities for peasants living on the agrarian reform settlements opened up in industry and mechanised agriculture. Landless workers who receive land through the agrarian reform process can theoretically access credit and technical assistance from the state to help build a house and for initial agricultural activities. However, credits are often paid late or not at all, technical assistance is unavailable or inadequate and the lack of infrastructure for the commercialisation of peasant products hampers the generation of income. Furthermore, the weather can lead to the loss of an entire season's harvest and therefore the loss of income. These are just some of the issues that drive (new) peasants into debt and indebted peasants no longer have access to credit from peasant credit programmes.[11]

Within the context of this lack of access to financial and adequate technical assistance, wage work in the sugarcane sector leads peasants

---

[9] Indigenous people on a reservation, interviews no. 7 and 8; and on Indigenous land, interview no. 9.
[10] This was mentioned various times during interviews that I was not allowed to record, or in informal conversations.
[11] Peasants on an agrarian reform settlement, interviews no. 10, 11 and 12; a member of the landless movement MST, Interview no. 13; a representative of Agraer, Interview no. 14; a representative of a rural labour union, Interview no. 15.

to remain on their land and finance necessary investments. This model existed before the expansion of sugarcane. Supplementing the work on their own land with (seasonal) work on large estates or other activities has been common. Nevertheless, sugarcane expansion allowed for a larger mass of people to become temporary wage workers. Various peasants made statements that provided evidence of this. They worked in the sugarcane sector for a period of time until they had saved enough money to (re)start their lives as peasants. Peasants from different agrarian reform settlements described temporary wage work as a way of remaining on the land.[12] The next section presents the changes that have occurred in subsistence work.

11.4.2 Subsistence Work: Land Prices and Access to Land

Land is the most important means of production in agricultural subsistence work. The expansion of the sugarcane industry had an impact on the availability, access and use of land due to increased land prices. Between 2002 and 2013, land prices in Mato Grosso do Sul increased by 586%, which was one of the highest increases in the country[13] (Sant'Anna et al. 2016b, p. 314). This surge in land prices can mostly be attributed to the expansion of the sugarcane sector. During the period which the prices rose to the greatest extent, the international commodity price for soybeans, the main agricultural product exported from Mato Grosso do Sul, was low[14]; therefore, the area planted with soybeans diminished between 2006 and 2012.[15]

The expansion of the sugarcane industry and the rising land prices occurred in an already conflictual context, where landless workers and Indigenous people had been (re)claiming land for decades (Brand et al.

---

[12] Peasants on an agrarian reform settlement, Interviews no. 10, 11 and 12; group discussion with peasants from different agrarian reform settlements, Interview no. 16.

[13] Between 2000 and 2005, the price rose from 2689 R$/ha to 4983 R$/ha for agricultural land and from 1644 R$/ha to 3220 R$/ha for pastureland (Gasques et al. 2008, S. 9–10).

[14] Three big landowners, Interviews no. 17, 18 and 19; a representative of a municipal Secretariat for Economic Development, Interview no. 20.

[15] Semagro, BDEWeb: see, http://bdeweb.semade.ms.gov.br/bdeweb/. Accessed 12 Nov 2020.

2008; Almeida 2003). Whereas the agrarian reform process had led to some land redistribution and some land to be demarcated as Indigenous,[16] land concentration remained high. In 2006, the number of large agrarian establishments with over 100 hectares (ha) only amounted to one-third of all agricultural properties, but they occupy 97% of the agricultural land.[17]

In general, redistributive agrarian politics are less likely in situations with increased land prices (Borras et al. 2011, p. 37). The number of new agrarian reform settlements has been decreasing since 2005 and new settlements have not been established since 2013. Certainly, this dynamic is not caused exclusively by the sugarcane expansion, but by the nationwide advancement of an export-led agribusiness combined with the political abandonment of the peasants and agrarian reform (Robles 2018). However, sugarcane expansion is one more reason for competing claims over land and it is particularly responsibly for the rise in land prices. All of this has changed access to land.

Indigenous land faces an even worse situation. When an area is approved as Indigenous, the previous land title becomes invalid. This means that the landowners do not receive any compensation for the land they lose. However, they do receive compensation for the *benfeitorias*, the cost of acquisition, creation or improvement of an asset such as a house or stable. When land prices were low, the costs of the *benfeitorias* exceeded the prices of the land. Since the land prices have increased, it has become unprofitable for landowners to merely receive compensation for their *benfeitorias,* as these now only cover a fraction of the land value.[18]

Big landowners, who usually plant soybean or sugarcane or raise cattle, resist demarcation via juridical measures. As soon as the demarcation process starts, they file an objection (Public Prosecution, Interview no.

---

[16] In demarcation processes, traditional Indigenous lands and their limits are supposed to be identified, declared, physically demarcated, homologated and registered. ISA (Instituto Socioambiental): see, https://pib.socioambiental.org/en/Demarcation. Accessed 29 Nov 2018.

[17] Own calculation based on Pavão (2005, p. 162) and Censos Agropecuários: see, sidra.ibge.gov.br. Accessed 13 May 2020.

[18] Interview with a big landowner who owns land that partly lies on identified Indigenous land, Interview no. 1 and with the public prosecutor MPF, Interview no. 21.

21). Landowners have even managed to have declarations of Indigenous land annulled. *Terras Indígenas* reports on their website that in 2015, 140 legal actions concerning Indigenous land demarcation in Mato Grosso do Sul went to the federal courts (Caliari 2016; Miotto 2018). The federal government demarcated the last Indigenous land in 2004.[19] Whereas the creation of new agrarian reform settlements was complicated for landless people/peasants, the Guarani-Kaiowá face the further problem that land that has been demarcated as Indigenous lands may be taken back. This demonstrates that the unequal access to land has not only been consolidated but exacerbated.

11.5 Discussion and Outlook: Labour Regimes in Sugarcane Industry Expansion

The main objective of this paper was to understand how the expanding production of biofuels as part of an emerging bioeconomy affect existing social inequalities in labour and land relations.

In Sects. 11.3 and 11.4, I demonstrated how the entanglements of global dynamics and national policies had an important influence on the local level. Without the dynamics of global land grabbing and the global "green development" discourse that arose in the wake of the Kyoto Protocol and COP21, the resurgence of the Brazilian sugarcane sector would have been difficult. The global green development discourse had an important impact on the mechanisation of the sugarcane harvest and attempts to reduce CO_2. In turn, this had a profound influence on local labour relations. The dynamics of global land grabbing and the increasingly strong sugarcane sector led to an expansion of the sugarcane industry and a subsequent surge in land prices. The higher land prices led to a deterioration in the access of Indigenous people to land and made future agrarian reform processes more difficult. This shows how important it is to move beyond the national level and to consider

---

[19]Terras Indígenas: see, https://terrasindigenas.org.br/pt-br#pesquisa. Accessed 25 Sept 2018.

global dynamics alongside national and local changes when examining social inequalities.

Section 11.4 demonstrated that the impact of sugarcane industry expansion differed depending on the social group in question. For the peasants in agrarian reform settlements, the change in the labour regime could best be described as an increased albeit temporal semi-proletarianisation. While the new work opportunities in the sugarcane sector provided broader access to wage labour and caused a wider semi-proletarianisation of the peasants, this process was not permanent. As soon as peasants had achieved a more stable income through their agriculture activities, they went back to being just peasants. Therefore, a process of re-peasantisation occurred in parallel to the process of semi-proletarianisation. This is why Bernstein (2010, pp. 33–34) describes the categories of "landless labour" and "small peasants" as fluid, because people move between these categories and occupy more than one at the same time. This also illustrates how peasants are able to benefit from their access to their land access simply through capital (wages), and this is also described by Ribot and Peluso (2003, pp. 160–171) in the *Theory of Access*.

In the case of the Guarani-Kaiowá, they were the principal labour force and, therefore, had access to wage labour before the mechanisation of the sugarcane industry. Even though the working conditions had often been slave-like, the sugarcane industry was one of the few possibilities that they had to secure subsistence (Abascal et al. 2016, p. 2), given that they had been denied access to their traditional land for decades (Brand et al. 2008). When the expansion and mechanisation occurred, they lost their access to wage work and therefore suffered a double exclusion: from land _and_ wage labour. Malnutrition, high child mortality and suicide rates (Abascal et al. 2016, pp. 1–2) show that securing livelihoods became very difficult. This phenomenon has been described as "expulsion" by Saskia Sassen:

> The past two decades have seen a sharp growth in the number of people, enterprises, and places expelled from the core social and economic orders of our time. […] The notion of expulsion takes us beyond the more

familiar idea of growing inequality as a way of capturing the pathologies of today's global capitalism. (2014, p. 1)

Sassen describes this dynamic as part of the deepening of capitalist relations. People once crucial to the development of capitalism stop being of value to the larger system. Natural resources, in this case land in of e.g. Latin America, are now viewed as become more important than the people who live on those lands as workers or consumers (ibid., p. 10).

These conclusions must be borne in mind when discussing the potential effects of the emerging bioeconomy. Even though the transition away from fossil fuels towards renewable energies is important, the impact on land and labour relations have to be considered, as is exemplified by the case of the Brazilian biofuel sector. This is especially relevant when the implementation of the bioeconomy includes the expansion of land-based biomass. Expansion dynamics do not necessarily lead to peasant expropriation, but can hinder more equal land distribution policies and may even destroy the livelihoods of Indigenous populations. Although dynamics of expansion such as these may have a positive impact on employment, a closer look reveals that the most vulnerable people (e.g. unskilled labourers) do not benefit from increased mechanisation and technologization. When reflecting on bioeconomy policies, a greater effort must be undertaken to consider how to avoid reproducing existing social inequalities and negatively impacting the most vulnerable population groups, such as unskilled workers, peasants and Indigenous populations.

List of Interviews quoted

To assure anonymity, names, gender, positions and detailed locations are omitted; in some cases, the name of the organisation is also withheld.

Interview no.	Institution/Organisation	Date and place
Interview no. 1	Big landowner	23/11/2017, Mato Grosso do Sul
Interview no. 2	University professor	20/11/2017, Mato Grosso do Sul
Interview no. 3	Person from an organic agriculture association	03/05/2018, Mato Grosso do Sul

(continued)

(continued)

Interview no.	Institution/Organisation	Date and place
Interview no. 4	Public Prosecution for Labour Rights (MPT)	13/06/2018, Mato Grosso do Sul
Interview no. 5	Representative of a labour union	09/05/2018, Mato Grosso do Sul
Interview no. 6	Representative of a labour union	20/06/2018, Mato Grosso do Sul
Interview no. 7	Indigenous person on a reservation	18/06/2018, Mato Grosso do Sul
Interview no. 8	Indigenous person on a reservation	18/06/2018, Mato Grosso do Sul
Interview no. 9	Indigenous person on Indigenous land	19/06/2018, Mato Grosso do Sul
Interview no. 10	Peasant on an agrarian reform settlement	16/06/2018, Mato Grosso do Sul
Interview no. 11	Peasant on an agrarian reform settlement	16/06/2018, Mato Grosso do Sul
Interview no. 12	Peasant on an agrarian reform settlement	16/06/2018, Mato Grosso do Sul
Interview no. 13	Member of the landless movement MST	22/11/2017, Mato Grosso do Sul
Interview no. 14	Representative of the agricultural development agency Agraer	10/11/2017, Mato Grosso do Sul
Interview no. 15	Representative of a rural labour union	16/11/2017, Mato Grosso do Sul
Interview no. 16	Group discussion with peasants from different agrarian reform settlements	20/04/2018, Mato Grosso do Sul
Interview no. 17	Big landowner	07/05/2018, Mato Grosso do Sul
Interview no. 18	Big landowner	07/05/2018, Mato Grosso do Sul
Interview no. 19	Representative of a rural union and big landowner	07/05/2018, Mato Grosso do Sul
Interview no. 20	Municipal Secretariat for Economic Development	07/05/2018, Mato Grosso do Sul
Interview no. 21	Public Prosecutor (MPF)	11/06/2018, Mato Grosso do Sul

References

Abascal, A., Karg, S., Kretschmer, R., Schaffrath-Rosario, A., & Schweikert, F. (2016). Brasilien: Der Kampf der Guarani-Kaiowá um Land und Würde. *FIAN Fact Sheet*, (1). http://www.fian.de/fileadmin/user_upload/dokumente/shop/Fallarbeit/2016-1_FS_Guarani_final_screen.pdf. Accessed 19 Nov 2020.

de Abreu, D., de Moraes, L.A., Nascimento, E.N., & Aparecida de Oliveira, R. (2009). Impacto social da mecanização da colheita de cana-de-açúcar. *Revista Brasileira de Medicina do Trabalho, Número Especial* (4, 5 and 6), 3–11.

de Almeida, R.A. (2003). *Identidade, distinção e territorialização: O processo de (re)criação camponesa no Mato Grosso do Sul*. Doctoral thesis. Universidade Estadual Paulista, Presidente Prudente. https://repositorio.unesp.br/bitstream/handle/11449/99830/almeida_ra_dr_prud.pdf?sequence=1&isAllowed=y. Accessed 8 May 2017.

Assunção, J., Pietracci, B., & Souza, P. (2016). Fueling Development: Sugarcane Expansion Impacts in Brazil. *Working Paper*. https://climatepolicyinitiative.org/wp-content/uploads/2016/07/Paper_Fueling_Development_Sugarcane_Expansion_Impacts_in_Brazil_Working_Paper_CPI.pdf.pdf. Accessed 12 March 2017.

Backhouse, M., Lorenzen, K., Lühmann, M., Puder, J., Rodríguez, F., & Tittor, A. (2017). Bioökonomie-Strategien im Vergleich: Gemeinsamkeiten, Widersprüche und Leerstellen. *Working Paper 1, Bioeconomy & Inequalities*, Jena. https://www.bioinequalities.uni-jena.de/sozbemedia/neu/2017-09-28+workingpaper+1.pdf. Accessed 10 Nov 2020.

Baer, W. (2014). *The Brazilian economy: Growth and development* (7th ed.). Boulder: Rienner.

Bernstein, H. (2010). *Class Dynamics of Agrarian change*. Halifax, N.S., Sterling: Fernwood Publishing; Kumarian Press.

Boatcă, M. (2016). *Global Inequalities Beyond Occidentalism*. London and New York: Routledge.

Bohrer, A. (2018). Intersectionality and Marxism: A Critical Historiography. *Historical Materialism, 26*, 46–74.

Borges, U., Freitag, H., Hurtienne, T., & Nitsch, M. (1983). *Proalcool: Analyse und Evaluierung des brasilianischen Biotreibstoffprogramms*. Freie Universität Berlin, Lateinamerika-Institut, Diskussionspapier 13. Berlin.

Borras, S.M., Franco, J.C., Kay, C., & Spoor, M. (2011). *Dynamics of Land Grabbing in Latin America and the Caribbean*. A paper prepared for and presented at the Latin America and Caribbean seminar: 'Dinámicas en el mercado de la tierra en América Latina y el Caribe', 14–15 November, FAO Regional Office, Santiago, Chile. https://www.tni.org/files/download/borras_franco_kay__spoor_land_grabs_in_latam__caribbean_nov_2011.pdf. Accessed 28 May 2020.

Brand, A.J., Ferreira, E.M.L., & de Azambuja, F. (2008). Os Kaiwá e Guarani e os processos de ocupação de seu território em Mato Grosso do Sul. In R.A. de Almeida (Ed.), *A questão agrária em Mato Gross do Sul: Uma visão multidisciplinar* (1st ed., pp. 27–52). Campo Grande: Universidade Federal do Mato Grosso do Sul. UFMS.

Brunner, J. (2017). *Die Verhandlungsmacht von Arbeiter*innen und Gewerkschaften in landwirtschaftlichen Transformationsprozessen: Eine Analyse des Zuckerrohrsektors im Bundesstaat São Paulo.* GLOCON Working Paper Series 6. http://www.land-conflicts.fu-berlin.de/_media_design/working-papers/WP-6-Jan_Online-final.pdf. Accessed 17 Jan 2018.

Caliari, T. (2016). Adeus, Guyraroká. Pública. https://apublica.org/2016/09/adeus-guyraroka/. Accessed 16 Oct 2018.

Damasceno, R., Chiavari, J., & Lopes, C.L. (2017). *Evolution of Land Rights in Rural Brazil: Framework for Understanding, Pathways for Improvement.* Rio de Janeiro. https://climatepolicyinitiative.org/wp-content/uploads/2017/06/Evolution_of_Land_Rights_In_Rural_Brazil_CPI_FinalEN.pdf. Accessed 8 Aug 2019.

Defante, L.R., Vilpoux, O.F., & Sauer, L. (2018). Rapid Expansion of Sugarcane Crop for Biofuels and Influence on Food Production in the First Producing Region of Brazil. *Food Policy*, *79*, 121–131.

de Rezende, S.B.A. (2014). Trabalho indígena no corte de cana-de-acúcar e atuação da Comissão Permanente no Estado de Mato Grosso do Sul. In M. Pauletti (Ed.), *Memorial da Comissção Permanente de Investigação e Fiscalização das Condições de Trabalho em Mato Grosso do Sul: 20 anos de história* (pp. 188–218). Campo Grande: Majupá.

Domingues, A.T. (2017). *A territorialização do capital canavieiro no Mato Mrosso do Sul: O caso da Bunge em Ponta Porã/MS.* Doctoral thesis. Universidade Federal da Grande Dourados, UFGD, Dourados.

Federici, S. (2012). *Aufstand aus der Küche: Reproduktionsarbeit im globalen Kapitalismus und die unvollendete feministische Revolution.* Münster: Ed. Assemblage.

Fernandes, B.M., Welch, C.A., & Gonçalves, E.C. (2010). Agrofuel Policies in Brazil: Paradigmatic and Territorial Disputes. *Journal of Peasant Studies*, *37*(4), 793–819.

Gasques, J.G., Bastos, E.T., Valdes, C., & USDA. (2008). Preços da Terra no Brasil. XLVI Congresso da Sociedade Brasileira de Economia, Administração e Sociologia Rural. Rio Branco, Acre.

Goven, J., & Pavone, V. (2015). The Bioeconomy as Political Project: A Polanyian Analysis. *Science, Technology & Human Values, 40*, 1–36.

Jesus, K.R.E., & Torquato, S.A. (2014). Evolução da mecanização da colheita de cana-de-açúcar em São Paulo: Uma reflexão a partir de dados do Protocolo Agroambiental. In UFRGS (Ed.), *Simpósio da Ciência do Agronegócio, 12./13. November 2014.* Porto Alegre: UFRGS.

Kreckel, R. (2004). *Politische Soziologie der sozialen Ungleichheit* (3rd ed.). Frankfurt and New York: Campus.

Li, T.M. (2017). The Price of Un/Freedom: Indonesia's Colonial and Contemporary Plantation Labor Regimes. *Comparative Studies in Society and History, 59*, 245–276.

Lorenzen, K. (2019). Sugarcane Industry Expansion and Changing Land and Labor Relations in Brazil: The Case of Mato Grosso do Sul 2000–2016. *Working Paper 9, Bioeconomy & Inequalities*, Jena. https://www.bioinequalities.uni-jena.de/sozbemedia/WorkingPaper9.pdf. Accessed 30 Oct 2019.

MCTIC. (2016). *Estratégia Nacional de Ciência, Tecnologia e Inovação 2016–2022.* Brasília. http://www.finep.gov.br/images/a-finep/Politica/16_03_2018_Estrategia_Nacional_de_Ciencia_Tecnologia_e_Inovacao_2016_2022.pdf. Accessed 16 Sep 2019.

Mies, M. (1988). Kapitalistische Entwicklung und Subsistenzproduktion: Landfrauen in Indien. In C. von Werlhof, M. Mies, & V. Bennholdt-Thomsen (Eds.), *Frauen, die letzte Kolonie: Zur Hausfrauisierung der Arbeit* (pp. 86–112). Reinbek: Rowohlt.

Miotto, T. (2018). Comunidade Guarani Kaiowá busca reverter no STF decisão que anulou demarcação. CIMI. https://cimi.org.br/2018/09/comunidade-guarani-kaiowa-busca-reverter-no-stf-decisao-que-anulou-demarcacao/. Accessed 15 March 2019.

Missio, F.J., & Vieira, R.M. (2015). A dinâmica econômica recente do Estado de Mato Grosso do Sul: Uma análise da composição regional e setorial. *Redes, 19*, 176.

Pauletti, M. (2014). *Memorial da Comissção Permanente de Investigação e Fiscalização das Condições de Trabalho em Mato Grosso do Sul: 20 anos de história.* Campo Grande: Majupá.

Pavão, E.D.S. (2005). *Formação, estrutura e dinâmica da economia do Mato Grosso do Sul no contexto das transformações da economia Brasileira.* Master thesis. Universidade Federal de Santa Catarina, Florianópolis.

Repórter Brasil (2017). *O que é trabalho escravo*. http://reporterbrasil.org.br/trabalho-escravo/. Accessed 27 Sep 2017.

Ribot, J.C., & Peluso, N.L. (2003). A Theory of Access. *Rural Sociology, 68*, 153–181.

Robles, W. (2018). Revisiting Agrarian Reform in Brazil, 1985–2016. *Journal of Developing Societies, 34*(1), 1–34.

Sant'Anna, A.C., Shanoyan, A., Bergtold, J.S., Caldas, M.M., & Granco, G. (2016a). Ethanol and Sugarcane Expansion in Brazil: What Is Fueling the Ethanol Industry? *International Food and Agribusiness Management Review, 19*, 163–182.

Sant'Anna, A.C., Granco, G., Bergtold, J.S., Caldas, Marcellus, M., Xia, T., Masi, P., et al. (2016b). Os desafios da expansão da cana-de-acúcar: A percepção de produtores e arrendatários de terras em Goiàs e Mato Grosso do Sul. In G.R. de Santos (Ed.), *Quarenta anos de etanol em larga escala no Brasil: desafios, crises e perspectivas* (pp. 113–143). Brasília: Ipea.

Sassen, S. (2014). *Expulsions: Brutality and Complexity in the Global Economy*. Cambridge: The Belknap Press of Harvard University Press.

Sauer, S., & Leite, S.P. (2011). *Agrarian Structure, Foreign Land Ownership, and Land Value in Brazil*. Paper presented at the International Conference on Global Land Grabbing 6–8 April 2011. http://citeseerx.ist.psu.edu/viewdoc/download?doi=10.1.1.459.1188&rep=rep1&type=pdf. Accessed 4 Aug 2017.

SEMAGRO. (2018). *Perfil Estatístico de Mato Grosso do Sul 2017*. http://www.semagro.ms.gov.br/wp-content/uploads/2018/12/Perfil-Estat%C3%ADstico-de-MS-2018.pdf. Accessed 06 May 2021.

Solga, H., Powell, J.J.W., & Berger, P.A. (Eds.). (2009). *Soziale Ungleichheit: Klassische Texte zur Sozialstrukturanalyse*. Frankfurt and New York: Campus.

Von Werlhof, C., Mies, M., & Bennholdt-Thomsen, V. (Eds.). (1988). *Frauen, die letzte Kolonie: Zur Hausfrauisierung der Arbeit*. Reinbek: Rowohlt.

Wilkinson, J. (2015). *The Brazilian Sugar Alcohol Sector in the Current National and International Conjuncture* (Paper prepared for Actionaid). Rio de Janeiro. http://actionaid.org.br/wp-content/files_mf/1493419528completo_sugar_cane_sector_ing.pdf. Accessed 28 May 2020.

Wilkinson, J., & Herrera, S. (2010). Biofuels in Brazil: Debates and Impacts. *The Journal of Peasant Studies, 37*, 749–768.

Yin, R.K. (2009). *Case Study Research: Design and Methods* (4th ed.). Los Angeles: Sage.

Open Access This chapter is licensed under the terms of the Creative Commons Attribution 4.0 International License (http://creativecommons.org/licenses/by/4.0/), which permits use, sharing, adaptation, distribution and reproduction in any medium or format, as long as you give appropriate credit to the original author(s) and the source, provide a link to the Creative Commons license and indicate if changes were made.

The images or other third party material in this chapter are included in the chapter's Creative Commons license, unless indicated otherwise in a credit line to the material. If material is not included in the chapter's Creative Commons license and your intended use is not permitted by statutory regulation or exceeds the permitted use, you will need to obtain permission directly from the copyright holder.

12

Territorial Changes Around Biodiesel: A Case Study of North-Western Argentina

Virginia Toledo López

12.1 Introduction

At the beginning of the new millennium, several biodiesel projects were announced in Argentina, driven mostly by the world market. However, a national law covering agrofuels was not passed until 2006.[1] The new law created a local consumer market for biodiesel and ethanol as of 2010. At this time, Argentina became the world's largest exporter and fourth-largest producer of biodiesel. Recently, this sector has experienced a further boom due to the promotion of the "bioeconomy", with

---

[1] The most widely used term is "biofuels"; however, this term has been questioned in environmental debates. In this article, I refer to them as "agrofuels", thus, avoiding the positive connotations inherent in the "bio" prefix. I define agrofuels as fuels based on flex crops that derive from industrial agriculture and are produced by agribusiness (Gras and Hernández 2016).

V. Toledo López (✉)
Institute of Studies of Social Development (INDES), University of Santiago de Estero (UNSE)-National Scientific and Technical Research Council (CONICET), Santiago de Estero, Argentina

biodiesel as the favoured product in this new area in terms of production volume and territorial expansion. The purpose of this chapter is to use a case study to understand the territorial changes related to biodiesel production in Argentina.[2]

The chapter begins with a brief description of the theoretical framework and the methodological approach used for the analysis. The next section considers the political and economic context of biodiesel production in Argentina. The sections that follow assess the territorial impacts of the agroindustrial frontier in north-western Argentina (NWA) and the specific territorial changes linked to biodiesel production within that area. This is done by focusing on the case of Santiago del Estero. Finally, some considerations of the ongoing process are drawn.

Social research in Argentina has already produced some noteworthy analyses of agrofuel public policies (Wehbe et al. 2008), their economic potential (Gorenstein and Gutman 2016; Rozemberg et al. 2009; Scheinkerman de Obschatko and Begenisic 2006; Carrizo et al. 2009; Chidiak et al. 2012; Dam et al. 2009), the effects of international specialization and trade (Lorenzo 2015). Studies have also been conducted into the environmental impact of agrofuels, and these also assess the energy balance of soya biodiesel (Donato et al. 2008; Iermanó and Sarandón 2009) and greenhouse gases (Hilbert and Galbusera 2011; Hilbert et al. 2012). Furthermore, Andersen et al. (2012) studied the relationship between land use and agrofuel based on different biomasses (soya, sunflower, jatropha). In addition, agrofuels have been considered in terms of extractivism (Toledo López 2013) and within a conceptual framework on energy and food sovereignty (Toledo López 2018). However, the research literature has not yet systematically analysed the social and environmental impact of biodiesel production from the perspective of political ecology (PE).

---

[2]This contribution is a synthesis of some findings from my doctoral thesis in social sciences, which I completed in 2016. The research was continued thanks to a postdoctoral fellowship from the National Council for Scientific and Technical Research, *Consejo Nacional de Investigaciones Científicas y Técnicas* (CONICET).

12.2 Theoretical and Methodological Framework

This work is situated within a broad political ecology (PE) perspective (Alimonda 2002, 2011; Martin and Larsimont 2016) and focuses on power within societal–nature relations, ecological distribution, environmental appropriation and valuation conflicts. Therefore, assuming that environmental impacts are unequally distributed amongst societies, social groups, communities and classes, I use the *territorial* approach from critical geography to elaborate on this perspective and to point to social practices that create a spatial distribution as a starting point for understanding the complexity around societal–nature relations, which are shaped by power (Harvey 1989, 2001; Santos 2000; Haesbaert 2007). Harvey argues that "the production of spatial organization" (2001, p. 327) involves the production of space and nature.[3] In this viewpoint, "territory" is a multiple, diverse and complex social construction that is shaped by simultaneous processes of domination, appropriation and resistance and expressed in both material and symbolic terms. In focusing on territorial changes around agrofuels, therefore, I refer to both the environmental and the social changes and the ways in which they are outlined by power relations. The focus on the material and symbolic dimensions of socio-ecological dynamics led me to consider the "valuation of nature", as promoted by different territorial agents, and the resulting conflicts (Harvey 1996, p. 150).

Environmental appropriation is considered a form of accumulation by dispossession that involves a wide range of processes.

> These include the commodification and privatization of land and the forceful expulsion of peasant populations; conversion of various forms of property rights – common, collective, state, etc. – into exclusive private property rights; suppression of alternative, indigenous, forms of production and consumption; colonial, neo-colonial and imperial process of appropriation of assets, including natural resources; monetization

---

[3]This implies that "material spatial practices", "representations of space" and "spaces of representation" of space (and time) are set by power relations (Harvey 1989, p. 220).

of exchange and taxation, particularly of land; slave trade; and usury, the national debt and ultimately the credit system. The state, with its monopoly of violence and definitions of legality, plays a crucial role in both backing and promoting these process [...]. Wholly new mechanisms of accumulation by dispossession have also opened up. (Harvey 2004, pp. 74–75)

In this sense, "bio" or agrofuels can be seen as a "new appropriation of nature" and the continuation of capitalism through projects with green ends (Fairhead et al. 2012).

This section first considers environmental transformations related to agribusiness expansion such as pesticide use, deforestation, loss of biodiversity, water contamination and the destruction of the material basis for life, especially in areas where both soya and biodiesel have been recently introduced. Second, this section also focuses on socio-political impacts, such as the construction of "development narratives" (Svampa and Antonelli 2009), that legitimize the practices of agribusiness. This finally leads to the concept of hegemony (Gramsci 2011; Gras and Hernández 2016). This concept helps provide a better understanding of the power dimension within the societal–nature relations identified by the case study (understood in the sense of Flyvbjerg 2011).

The methodological approach includes a combination of a literature review, statistics and reports by official Argentinian institutions, media data and in-depth, field-based research. Primary sources were collected mostly in 2012, when (participatory) observations and about 30 semi-structured interviews were conducted with employees and former staff from a biodiesel factory, managers and administrative staff, local and provincial public officials, technicians, key informants, members of educational institutions and non-governmental organizations, neighbours and small farmers from the area.

12.3 Agrofuels Production in Argentina

To understand the agrofuels boom and its territorial implications, I first consider the conditions that made biodiesel production in Argentina possible. A crucial moment in this process was when permission was

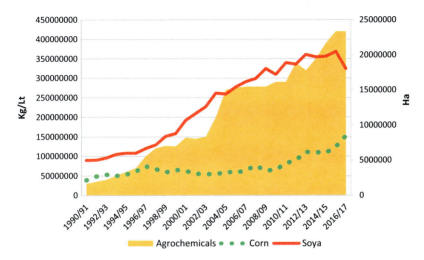

Fig. 12.1 Agrochemicals use (kg/lt) and soya Farmland (ha). 1990/1991–2016/2017 (*Source* Own elaboration, adapted from *Sistema de Datos Abiertos de la Secretaría de Agroindustria*, https://datos.agroindustria.gob.ar/dataset/est imaciones-agricolas and *Naturaleza de derechos* 2019)

granted to produce and trade glyphosate-tolerant soya (a genetically modified organism—GMO) in 1996 (Teubal 2009). In 1980, less than one million hectares (ha) of land was being used for soya farming; by 2012/2013, this had risen to 20 million ha out of a total of 34 million ha of agricultural land in the country.[4] The introduction of GMOs led to an intensification and expansion of industrial agriculture—driven by soya production (Gras and Hernández 2013, 2016).

As a result, soya became the main annual crop, in terms of both land use and production levels. The approval of further GMOs extended this logic to other products (today, more than 60 GMOs are authorized in Argentina) and positioned Argentina as the country with the third-largest level of land used for GMOs, or 12.5% of the world's farmland. This also led to an increased use of pesticides in agriculture. As Fig. 12.1 shows, the use of agrochemicals has increased from 30 million lt/kg in 1990 to 525 million lt/kg in 2018. In recent years, Argentina has become the

---

[4]See, https://datos.agroindustria.gob.ar/dataset/estimaciones-agricolas. Accessed 29 Oct 2019.

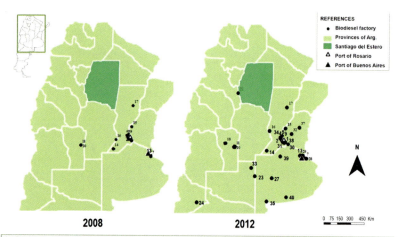

Fig. 12.2 Biodiesel agroindustry in Argentina. Location in 2008 and 2012 (*Source* Own elaboration, adapted from *Secretaría de Energía*. See, https://www.argentina.gob.ar/produccion/energia/. Accessed 12 May 2015)

focus of growing conflicts and controversies over environmental issues and the health consequences associated with agrochemical use.

The production of oilseed, led by soya (93% of the total), is the country's main export (amounting to almost 30% of total exports). In particular, the oil industry is export-oriented, and its main input is soya, with more than 90% of production aimed at the world market (MH[5] 2017). As such, it could easily be adapted to biodiesel production. The possibility of adding value to soya grain by turning it into fuel was attractive for sectors that saw "agriculture as a business" (Gras and Hernández 2013).

The biodiesel agroindustry in Argentina also contributes to economic and territorial concentrations. As Fig. 12.2 shows, the location of the first factories, close to the ports of Rosario and Buenos Aires, demonstrates that the agroindustry is focused on exports and the territorial

---

[5]Ministerio de Hacienda – Ministry of Finance.

concentration of biodiesel. 80% of Argentina's biodiesel production takes place close to this area. There are currently 50 active biodiesel companies, but the 8 most important (all of which are primarily oil-focused agroindustrial) companies produce almost 80% of the country's biodiesel and mainly target the world market (MH 2017). In particular, the large-scale biodiesel factories profited from the introduction of a national quota for biodiesel (see below): the four biggest enterprises alone supplied 40% of the biodiesel quota.

Biodiesel production rose from the least important source of agrofuels (around 711,864 tonnes) in 2008 to around 2.5 million tonnes (t) in 2012 (Fig. 12.3). Before 2010, all Argentinian biodiesel was destined for export. The rise in its production is mostly explained by increased demand from the European Union, which was initially the only destination of Argentinian biodiesel. As such, the supply of agrofuels in Argentina, which was strongly driven by the production of biodiesel, was stimulated by the world market and the promising exchange rate available as of 2002 (Toledo López 2013). In a global context, which has been

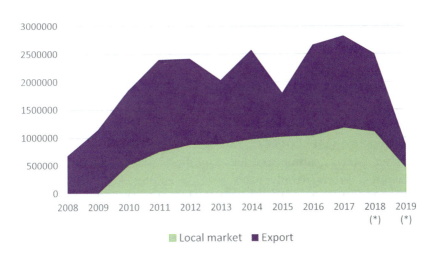

Fig. 12.3 Main destinations of Argentinian Biodiesel (t). Provisional data (*Source* Secretaría de Energía. See, https://www.argentina.gob.ar. Accessed 29 Oct 2019)

described as the "commodities consensus" (Svampa 2012) and the "neo-developmental stage" of the Argentinian accumulation regime (Féliz and López 2012), agrofuels were promoted through a green-friendly narrative that emphasized the industrialization of "existing raw materials" (Toledo López 2017). During this period, many agrofuel factories opened in the country, mostly located close to the area used for soya production and to the ports of Buenos Aires and Rosario, as Fig. 12.2 shows.

It was not only in 2006 that a national law covering agrofuels was passed: the "Biofuels Law" 26093, established a local consumer market for biodiesel and ethanol. It also created a national quota stipulating that petrol and gasoil had to contain a minimum blend of 5% of agrofuels as of 2010 and set a promotional regime for enterprises that became biofuel providers. In July 2010, the quota was increased to 10%. Since 2006, Law 26190 (updated in 2015 by Law 27191) has promoted agrofuels as renewable energy sources of electricity generation. Thus, agrofuels have taken on a leading role, not only in the transport sector, but also as part of an energy diversification policy. Nevertheless, with less than 10% of the national energy matrix, renewable sources continue to occupy a marginal space in Argentina's energy supply.

By 2010, Argentina was the world's fourth-largest biodiesel producer and the world's largest exporter. In terms of agricultural land use, estimates suggest that about a quarter of the soya crop is used for fuel production in the country.[6] In 2016, official data shows that Argentina produced 2.65 million t of biodiesel from soya oil, of which 39% was used for the national quota, with the remaining one and a half million t being exported (MH 2017).

Under Macri's administration,[7] agrofuels acquired new dynamics based on a green-friendly narrative that emphasized the benefits of renewable energy sources in solving the energy crisis as part of a liberalized macroeconomic programme (Seoane 2017; Varesi 2016; Féliz 2016; Toledo López 2018). In this context, the government gradually introduced the bioeconomy narrative (see Tittor in this volume) as a

---

[6] Ámbito (2011, April 19). See, https://www.ambito.com/edicion-impresa/la-produccion-biodiesel-demando-un-26-la-cosecha-n3678440. Accessed 1 Feb 2020.
[7] President of Argentina between 2015 and 2019.

new perspective, and in 2017, the Ministries of Science, Agroindustry and Production signed an agreement to promote it (CONVE-2017-12130310).[8] Biodiesel is still the main product in this new bioeconomy consensus, and it is viewed as the path to achieving sustainable development and shifting capitalism to "green" or "bio" production patterns.

The next two sections assess the general territorial impacts of the agroindustrial frontier in north-western Argentina and the specific impact of biodiesel production by one newly built factory within that area.

12.4 Territorial Changes Due to Agribusiness in North-Western Argentina

From a socio-economic perspective, the agricultural processes associated with agribusiness expansion involved deep changes to Argentinian rural areas. These changes included intensification by mechanization, the use of new technologies, the exclusion of the rural workforce, the expansion of flex crops, increased production for export and the deregulation of diverse productive activities in rural areas (Gras and Hernández 2016). Furthermore, economic liberalization led to increased economic and land concentration (Bisang and Gutman 2005; Giarracca and Teubal 2013; Gras and Hernández 2013, 2016).[9] This process has strengthened vertical integration, leading to the creation of value chains that are mostly controlled by a small number of enterprises (Teubal 2006). Territorial changes were particularly visible in NWA, where this process stimulated the growth of "flex crops" (Gras and Hernández 2013, 2016) and changed the traditional role of these northern rural territories from suppliers of local markets to suppliers of the world market (Gorenstein et al. 2011).

---

[8] See, http://www.bioeconomia.mincyt.gob.ar/wp-content/uploads/2017/07/Convenio-bioeconomia.pdf. Accessed 15 March 2020.
[9] According to census data, between 1960 and 1988, 51,000 farms disappeared in Argentina: 1800 per year. Between 1988 and 2002, this trend intensified to 6263 farms per year. Those that disappeared were smaller than 200 ha. In contrast, farms larger than 500 ha increased, particularly those in the stratum from 1000 to 2500 ha (Teubal 2006, p. 81).

Located in NWA, the province of Santiago del Estero is known for its biological and cultural diversity, as well as being home to one of the largest numbers of peasants in Argentina, and a huge area of native forest. It is also characterized by the lowest per capita income and the highest rate of poverty, in addition to a traditional migratory flow towards the centre of the country (Neiman 2009). Primary activities such as small-scale farming, forestry etc., contributed greatly to Santiago del Estero's economy, but this has changed with a growing move towards agribusiness (initially marked by soya production, but more recently by corn production). The expansion of the agribusiness model has led many traditional activities to disappear, relocate or intensify, as is the case with livestock production (Pengue 2017). As such, bioeconomy expansion has led to the enhancement of some economic variables, but has not necessarily led to improvements in people's living conditions: in 2017, the Sustainable Development Index, which measures economic growth, social inclusion and environmental sustainability, rated the province last in the country, with an index score of just 0.31 (PNUD—Programa Naciones Unidas para el Desarrollo 2017).

Despite the overall drop in the size of rural populations in Argentina, Santiago del Estero has one of the highest numbers of rural inhabitants (31.3% according to CNPNyV 2010-INDEC). However, statistics show a high level of precarity in land tenure patterns: census data demonstrate that half of the farms in the area are unable to extend their territory because they lack property titles or due to unclear boundaries (CNA—Censo Nacional Agropecuario 2002). This illustrates the social vulnerability of peasants, family farmers and smallholders, who face threats to their property rights and (dis)possession of the lands on which they live.[10] Inhabitants of rural areas, such as those in small areas of forest, are currently surrounded by large monocultures. Moreover, these people lack protection from violence and have no guaranteed human rights, as the growing conflicts around pesticide use, the privatization of land, the expulsion of peasants from their land, the conversion of various

---

[10]Argentinian law recognizes the rights of people to own land that possess through traditional occupation.

forms of property rights into private property and the suppression of native forms of production and consumption reveal.

The advancement of the agroindustrial frontier has involved an increase in grain and oilseed production, with changes to both the environmental and the social systems. According to census data, in 1988, 16% of farmland in the province was used to produce soya; in 2002, this had reached 38%; and in 2015, 58%, in other words, 980,572 ha of land are used to produce 2,498,134 t of soya (3.1% of the total Argentinian crop). In addition, recent years have seen an increase in corn production (MHFP and PEPD[11] 2016).

Several studies relate agribusiness expansion to higher levels of deforestation and loss of biodiversity, especially in the native forests of the Chaco, which is the largest dry forest in South America (Pengue 2017; Morello and Rodríguez 2009; Langbehn and Schmidt 2017). Regarding forests and their use, data from the corresponding monitoring unit (UMSEF 2016) show that 245,653 ha of native forest were cut down between 2009 and 2011 in Santiago del Estero; an additional 255,655 ha were cut down between 2011 and 2015. Thus, deforestation has destroyed the ecosystem in the Chaco where agribusiness has expanded. The historical coexistence of many smallholders and peasant communities in forest areas was built on a form of economic activity that values the ecosystem; this contrasts widely with the practices of agribusiness. Agribusiness expansion has particularly affected the continuity of small farmers and peasant forms of production, and their resistance to this situation is evident in the increasing conflicts over land (Slutzky 2005; Domínguez and Sabatino 2006; REDAF—Red Agroforestal Chaco 2013). In addition, some of these conflicts have been interpreted as accumulation by dispossession and land grabbing (Cáceres 2015; Gras 2017; Gras and Zorzoli 2019).

Thus, the expansion of the agroindustrial frontier in Santiago del Estero, which resulted in the production of soya for biodiesel, led to deep territorial transformations in terms of ecological distribution and environmental appropriation. These processes have had consequences

---

[11] Ministerio de Hacienda y Finanzas Públicas & Secretaría de Política Económica y Planificación del Desarrollo which are the Ministry of Finance and Public Spending & Secretary of Economic Policy and Development Planning.

that involve economic, environmental and social aspects, which I define as territorial changes. Recently, the biodiesel boom has encouraged this capital-intensive mode of production, the consequences of which include the commodification and privatization of land, the forceful expulsion of peasant communities; the conversion of various forms of property rights into private property; the suppression of alternative forms of production and consumption; and the degradation of environmental commons. Additionally, the biodiesel industry has developed specific elements of territorial control concerning both the material and the symbolic dimensions, including the development of new mechanisms of accumulation by dispossession, and connecting their businesses to environmental discourse. Biodiesel production has encouraged this process by strengthening the presence of agribusiness in the area in many ways, and this has added new dimensions to the territorial changes that it causes. This will be explained further in the next section.

12.5 Biodiesel and the Impacts of Agroindustry in Santiago Del Estero

In 2009, a mega biodiesel production venture, Viluco, with capacity to grind one million t of soya beans and to generate 200,000 t of fuel per year, started operation in Santiago del Estero (no. 22, Figure 12.2). It was the first agroindustrial biodiesel plant to be located in NWA, where the economic, social and environmental transformations caused by the expanding agroindustrial frontier are still underway. Indeed, the provincial government had advocated the growth of this "value chain", and it promoted the industrialization of the province with "Industrial Promotion" Provincial Law 6.750. The law was passed in 2005, and it was accompanied by infrastructure development in order to attract investment. Thus, government action strengthened rural actors by providing greater levels of capitalization (for Viluco). This was in line with the expansion of agribusiness and the "development narrative" that stressed the goal of "adding value" to raw materials. This was foreseen as part of the "neo-developmental stage" of the Argentinian accumulation regime (Féliz and López 2012; Svampa and Antonelli 2009). For instance, locating the plant in NWA meant a change in the spatial trend of biodiesel agroindustry. I view this as a signal of a progressing agribusiness frontier. Viluco also explicitly aimed to take advantage of the "lack" of a biodiesel plant in the region (Toledo López 2016).

In this sense, the case study provides more information about territorial changes associated with agribusiness, focusing on biodiesel production and how power relations affect ecological distribution in two ways. First, as the testimony of a rural priest from Santiago del Estero shows, the presence of a biodiesel company—as a key agent of agribusiness expansion in the area—contributed to the decline of communal forms of life:

> the reality of the peasant communities in our area and the difficulties faced by many of our brothers and sisters in the countryside, especially in terms of health, production, animal breeding and in the fields, are due to aerial and land spraying by large companies that are dedicated to soya bean plantations here in our area. There are large companies such as Viluco, part of the Lucci Group from Tucuman […], to name a few of the big businesses that have thousands and thousands of hectares of soya bean and who periodically spray their soya fields, especially with light aircraft… and peasant communities often find themselves isolated by these farms and are facing health problems. (local priest, Interview no. 1)

Secondly, some irregularities in effluent management are demonstrative of inequalities in the distribution, use and access to water sources in this dry ecoregion in which water is scarce. This led to controversy over the contamination of water related to biodiesel production, which can be understood in terms of environmental appropriation. The first production of biodiesel in the town of Frias occurred in June 2010 (as part of the national quota) and in the beginning of the plant released untreated effluents. The delay in considering the treatment of liquid wastes from industrial processes was said to be justified because the "crown technology" they used was promoted by the company as "zero effluent".[12] This led state officials to approve the obligatory environmental impact study without considering these wastes: "when they presented the Environmental Impact Valuation it was approved; it was not until they [Viluco] were up and running that we realized that effluents were being produced" (environmental director, Interview no. 2).

In search of a "solution", the effluents were routed through canals that had been created as part of flood prevention measures to channel water to the (dry) Albigasta River. This led to a conflict because the "solution" affected the local area. The first formal complaints about biodiesel liquid effluents were made to the ombudsman of the town of Frias in the beginning of 2011. People in the town complained about a bad smell and pointed to the death of horses and goats that had dunk water from the canal. At the same time, the local population underlined the lack of water in the area. However, state officials and the company questioned the toxicity of liquid effluents and the dominant opinion was that the effluents were biodegradable. The following statement is exemplary of this position: "Which products are washed out during the production process? Oil, some soya oil, which is biodegradable, and traces of flour" (secretary of production, Frias Municipality, Interview no. 3). This assertion ignores the fact that the process through which biodiesel is obtained requires toxic agents such as methanol (Sorichetti and Romano 2012). Asked about the lack of foresight when it came to effluents, the ombudsman stated: "I asked them the same question and they told me exactly the same thing: 'We'll deal with it after the company starts operating'" (Frias ombudsman, Interview no. 4). In summary, the case study shows that biodiesel production involves a risk of water pollution and that this can affect the living conditions of local communities in different ways.

Third, the way Viluco implemented its business, ignoring environmental law and social and ecological impacts, reveals the power

---

[12]El Liberal (2010, Aug 11). See, http://biodiesel.com.ar/3667/ag-energy-el-primer-biodiesel-pro ducido-en-la-localiad-de-frias-en-santiago-del-estero-sale-al-mercado-nacional. Accessed 18 July 2012.

dimension of societal–nature relations and how it influences ecological distribution as well as environmental appropriation and valuation practices. Indeed, state officials call the company "a source of pride for Santiago del Estero and for the town of Frias" (school director, Interviews no. 6 and 7). Once the controversy around the water contamination became public, the company signed a "Clean Production Agreement" with the provincial government. As part of the agreement, the company committed itself to building an effluent treatment in three stages and to preparing an "Environmental Management Plan". State officials used their speeches to underscore the company's compliance with the plan as a way of safeguarding its prestige and image and, thus, ensuring that the development narrative remained linked to the venture. In this sense, they highlighted that some environmental aims must be "sacrificed" in chase of "development", differentiating between what is possible and what is "desirable" (environmental director of Santiago del Estero, Interview no. 2). Therefore, once again the "development narrative" managed to obfuscate the negative social and ecological effects of biodiesel production. However, the solution to the controversy over effluents did not satisfy some locals, who viewed their way of life as under threat. Nevertheless, the current distribution of power places the company in a superior position to the demands made by the local population. The problem increased because the company continued to release effluents while the treatment pools were being constructed. This issue remained unresolved and the conflict only diminished when the company announced its closure in April 2019.[13] At the same time, the authority displayed by the company was also evident in the huge level of vulnerability faced by workers due to the "massive layoffs", suspensions (workers, Interview no. 5) and the lack of agencies able to guarantee their rights.

Finally, the case study reveals that pedagogical practices were carried out by the company in public and private schools in the town of Frias as a way of spreading its "development narrative" and encouraging young people to consider a job at the company. As a sign of its corporate social responsibility, the executive director of the company's foundation put

---

[13]Clarín (2019, April 26). See, https://www.clarin.com/rural/viluco-cierra-fabrica-biodiesel_0_kBkFhpQlv.html. Accessed 29 Oct 2019.

forward a "Community Integration Plan" to the local authorities that included "values tutoring", and courses and training for school teachers as "volunteers" (schools' director, Interview no. 6 and 7). These kinds of practices led the company to play an educational and moral role in the town, through which it reinforced existing power relations, a process that can be understood as building hegemony (Gramsci 2011), and that is related to appropriation by dispossession.

12.6 Conclusions

The biodiesel agroindustry in Argentina has contributed to economic concentration and territorial deterioration, and this has reinforced the impact of agribusiness. The land used for soya production has expanded dramatically in NWA, threatening native forests, grassland and fragile ecosystems, and displacing family farming, traditional agriculture, peasants' communities and indigenous people. The announcement that the biodiesel project was to be built in this area implied a new stage in agribusiness expansion. Nevertheless, it did not generate any major resistance at the local level, despite the numerous conflicts over land connected to the expansion of the industrial agricultural frontier. The case study of biodiesel production in Santiago del Estero reveals new impacts on the area linked to the degradation of local and communal ways of life, labour precarization, water pollution and the appropriation of common goods in the territory due to agribusiness expansion. In terms of societal–nature relations, the territory perspective shows that biodiesel is connected to processes of accumulation by dispossession, which are also linked to agribusiness territoriality. Moreover, the case study shows how biodiesel production involves shifting capitalism to "green" or "bio" productions. This takes place through the "valuation of nature", as promoted by agribusiness agents, the commodification and privatization of land and the forceful expulsion of peasant populations, and the deterioration of the material basis for life (including deforestation and the contamination of water, soil and air). The company's high level of symbolic power is analogous to the "development narrative" promoted by the state (Toledo López 2016), which also helped state authorities exercise territorial control.

The state plays a crucial role by promoting these (export-oriented) activities that reinforce the reprimarization of productive structures with "green" or "bio" narratives and modernization discourses. The dominant discourse of ecological modernization in the case of biodiesel production in Santiago del Estero, reinforced by the bioeconomy narrative, shows that the relationship between nature and society continues to be shaped by money (Harvey 1996, p. 150). In this perspective, nature is seen as a "resource" within capitalism and a paradigm that views economic and environmental goals as in opposition to one another. In this way, the study also illustrates the "valuation of nature" promoted by different territorial agents and the resulting conflicts (ibid.). This, in turn, affects the ecological distribution and the use of common goods for the benefit of profit-making and businesses.

It is also important to recognize how the distribution of power influences the government's environmental action. For instance, the company's environmental practices reveal the value placed on the symbolic dimension of territorial accumulation processes, which can also be defined as ecological appropriation, or a mode of green grabbing (Fairhead et al. 2012). In this perspective, considering the bioeconomy as a development narrative enables us to visualize the intrinsic valuation conflict underlying this type of territorial process, as emphasized within PE literature. Finally, biodiesel appears as a crucial product in this new bioeconomy consensus and is viewed as a manner of recycling capitalism through green (neo)developmental narratives.

List of Interviews quoted

Interview no.	Institution/oganization	Date and place
Interview no. 1	Local priest	17/05/2019, La voz de la Pacha, Capital—Santiago del Estero
Interview no. 2	Environmental director of the province of Santiago del Estero	19/03/2012, Capital—Santiago del Estero

(continued)

(continued)

Interview no.	Institution/oganization	Date and place
Interview no. 3	Secretary of production of Frias Municipality	20/03/2012, Frias—Santiago del Estero
Interview no. 4	Frias town ombudsman	26/03/2012, Frias—Santiago del Estero
Interview no. 5	Workers from the agroindustry	26/03/2012, Frias—Santiago del Estero
Interview no. 6	Director of school no. 1	27/03/2012, Frias—Santiago del Estero
Interview no. 7	Director of school no. 2	27/03/2012, Frias—Santiago del Estero

References

Alimonda, H. (2002). *Ecología política. Naturaleza, sociedad y utopía*. Buenos Aires: CLACSO.

Alimonda, H. (2011). *La Naturaleza colonizada. Ecología política y minería en América Latina*. Buenos Aires: CLACSO.

Andersen, F., Iturmendi, F., Espinosa, S., & Diaz, M. S. (2012). Optimal Dsign and Planning of Biodiesel Supply Chain with Land Competition. *Computers & Chemical Engineering, 47*, 170–182.

Bisang, R., & Gutman, G. (2005). Acumulación y Tramas Agroalimentarias en América Latina. *Revista de la CEPAL, 87*, 115–129.

Cáceres, D.M. (2015). Accumulation by Dispossession and Socio-Environmental Conflicts Caused by the Expansion of Agribusiness in Argentina. *Journal of Agrarian Change, 15*(1), 116–147.

Carrizo, S., Guibert, M., & Berdolini, J. (2009). Actores y mercados de los biocombustibles argentinos: entre incertidumbre y diversificación. Presentation 12th Encuentro de Geógrafos de América Latina (EGAL), 3–7 April 2009, Montevideo, Uruguay.

CNA (2002). Definiciones censales y metodología de relevamiento. INDEC.

Chidiak, M., Rozemberg, R., Fillipello, C., Gutman, V., Rozenwurcel, G., & Affranchino, M. (2012). Sostenibilidad de biocombustibles e indicadores GBEP: un análisis de su relevancia y aplicabilidad en Argentina.*Documento IDEAS, 11.*

Dam, J., Faaij, A.P.C., Hilbert, J., Petruzzi, H., & Turkenburg, W.C. (2009). Large-Scale Bioenergy Production from Soya Beans and Switchgrass in Argentina. Part A. Potential and Economic Feasibility for National and International Markets. *Renewable and Sustainable Energy Reviews, 13*, 1710–1733.

Domínguez, D., & Sabatino, P. (2006). Con la soja al cuello: crónica de un país hambriento productor de divisas. In H. Alimonda (Ed.), *Los tormentos de la materia. Aportes para una ecología política latinoamericana* (pp. 249–274). Buenos Aires: CLACSO.

Donato, L.B., Huerga, I.R., & Hilbert, J.A. (2008). Balance Energético de la producción de biodiesel a partir de soja en la República Argentina. INTA. https://inta.gob.ar/sites/default/files/script-tmp-bc-inf-08-08_balance_energetico.pdf. Accessed 15 May 2020.

Fairhead, J., Leach, M., & Scoones, I. (2012). Green Grabbing: A New Appropriation of Nature? *Journal of Peasant Studies, 2*(39), 237–261.

Féliz, M., & López, E. (2012). *Proyecto neodesarrollista en la Argentina ¿Modelo nacional popular o nueva etapa del desarrollo capitalista?* Buenos Aires: Herramienta-El colectivo.

Féliz, M. (2016). Argentina: cambió el gobierno, ¿cambió el proyecto hegemónico? *Herramienta, 58.*

Flyvbjerg, B. (2011). Case Study. In N. Denzin & Y. Lincoln (Eds.), *The Sage Handbook of Qualitative Research* (pp. 301–316). Thousand Oaks: Sage.

Harvey, D. (1989). *The Condition of Postmodernity. An Enquiry into the Origins of Cultural Change.* Oxford: Blackwell.

Harvey, D. (1996). *Justice, Nature, and the Geography of Difference.* Oxford: Blackwell.

Harvey, D. (2001). *Spaces of Capital. Towards a Critical Geography.* Edinburg: Routledge.

Harvey, D. (2004). The New Imperialism. *Socialist Register, 40*, 63–87.

Haesbaert, R. (2007). Território e Multiterritorialidade: um debate. *GEOgraphia, 9*(17), 19–45.

Giarracca, N., & Teubal, M. (2013). *El campo argentino en la encrucijada.* Buenos Aires: Alianza Editorial.

Gorenstein, S., & Gutman, G. (2016). Nuevos debates sobre acumulación, desarrollo y territorio: Clusters tecnológicos en la periferia. *Petróleo, Royalties & Região, 51*(3), 8–17.

Gorenstein, S., Schorr, M., & Soler, G. (2011). Dinámicas cambiantes de los complejos productivos en el norte argentino: los casos del tabaco, yerba mate y la soja. Un enfoque estilizado. *Revista interdisciplinaria de estudios agrarios, 34*, 5–33.

Gramsci, A. (2011): *Antología*. Buenos Aires: Siglo XXI.

Gras, C. (2017). Expansión sojera y acaparamiento de tierras en Argentina. *Desarrollo económico, 57*(221), 149–163.

Gras, C., & Zorzoli, F. (2019). Ciclos de acaparamiento de tierra y procesos de diferenciación agraria en el noroeste de Argentina. *Trabajo y Sociedad, 33*, 129–151.

Gras, C., & Hernández, V. (2016). Hegemony, Technological Innovation and Corporate Identities: 50 Years of Agricultural Revolutions in Argentina. *Journal of Agrarian Change, 16*(4), 675–683.

Gras, C., & Hernández, V. (2013). *El agro como negocio. Producción, sociedad y territorios en la globalización*. Buenos Aires: Biblos.

Hilbert, J.A., & Galbusera, S. (2011). Análisis de emisiones. Producción de biodiesel – Ag Energy. INTA. http://inta.gob.ar/documentos/analisis-de-emisiones-produccion-de-biodiesel-2013-ag-energy/. Accessed 1 March 2014.

Hilbert, J.A., Sbarra, R., & López-Amorós, M. (2012). Producción de biodiesel a partir de aceite de soja. Contexto y Evolución Reciente. INTA. http://inta.gob.ar/sites/default/files/script-tmp-inta_biodiesel_de_aceite_de_soja_en_argentina.pdf. Accessed 1 March 2014.

Iermanó, M.J., & Sarandón, S.J. (2009). Is It Sustainable the Production of Biofuels in Large Scale? The Case of Biodiesel in Argentina. *Revista Brasileira de Agroecologia, 4*(1), 4–17.

Langbehn, L., & Schmidt, M. (2017). Bosques y extractivismo en la Argentina. *Voces en el Fenix, 60*(8), 88–95.

Lorenzo, C. (2015). Domestic order and Argentinian's foreing policy: The issue of the biofuels. *Austral: Brazilian Journal of Strategy & International Relations, 4*(7). https://doi.org/10.22456/2238-6912.42443.

MH (2017). Informes cadena de valor. *Oleaginosas, 2*(29). https://www.argentina.gob.ar/sites/default/files/sspmicro_cadena_de_valor_oleaginosa.pdf. Accessed 29 Oct 2019.

MHFP & PEPD (2016). Informes productivos provinciales: Santiago del Estero. 1(9).

Martin, F., & Larsimont, R. (2016). ¿Es posible una ecología cosmo-política?: Notas hacia la desregionalización de las ecologías políticas. *Polis, 15*(45), 273–290.

Morello, J., & Rodríguez, A. (2009). *El Chaco sin bosques: la Pampa o el desierto del futuro.* Buenos Aires: Orientación Gráfica Editora.

Naturaleza de derechos (2019). "Collage de la depreciación humana". Investigación naturaleza de derechos. https://drive.google.com/file/d/1vb32EJJ7Trm0mP7EZrVZCHZT9T3lp7yu/view. Accessed 4 Nov 2019.

Neiman, G. (2009). *Estudio exploratorio y propuesta metodológica sobre trabajadores agrarios temporarios.* Buenos Aires: Ministerio de Economía y Producción, Secretaría de Agricultura Ganadería, Pesca y Alimentos. Proyecto de Desarrollo de Pequeños Productores Agropecuarios (PROINDER).

Pengue, W. (2017). *Cultivos Transgénicos ¿hacia dónde fuimos? veinte años después: la soja argentina 1996–2016.* Buenos Aires, Santiago de Chile: Fundación Heinrich Böll, GEPAMA.

PNUD (2017). Informe Nacional sobre Desarrollo Humano 2017. Información para el desarrollo sostenible: Argentina y la Agenda 2030. Buenos Aires.

REDAF (2013). Conflictos sobre tenencia de tierra y ambientales en la región del Chaco argentino: 3º Informe. Reconquista (Arg.): REDAF. http://redaf.org.ar/wp-content/uploads/2013/07/3informeconflictos_observatorioredaf.pdf. Accessed 29 Oct 2019.

Rozemberg, R., Saslavsky, D., & Svarzman, G. (2009). La Industria de Biocombustibles en Argentina. In López, A. (Coord.). *La Industria de Biocombustibles en el Mercosur*, Serie Red Mercosur (15).

Santos, M. (2000): *La naturaleza del espacio. Técnica y tiempo. Razón y emoción.* Barcelona: Ariel.

Scheinkerman de Obschatko, E., & Begenisic, F. (2006). *Perspectivas de los biocombustibles en la Argentina y Brasil.* Buenos Aires: IICA.

Seoane, J. (2017). El tratamiento neoliberal de la cuestión ambiental. http://www.opsur.org.ar/blog/2017/09/12/el-tratamiento-neoliberal-de-la-cuestion-ambiental/. Accessed 1 April 2019.

Slutzky, D. (2005). Los conflictos por la tierra en un área de expansión agropecuaria del NOA. La situación de los pequeños productores y los pueblos originarios. *Revista Interdisciplinaria de Estudios Agrarios, 23*, 59–100.

Sorichetti, P.A., & Romano, S.D. (2012). Uso de agua en la purificación de biodiesel: optimización mediante el control de propiedades eléctricas de efluentes. VII Congreso de Medio Ambiente, AUGM, 22–24 May, UNLP,

La Plata. http://sedici.unlp.edu.ar/bitstream/handle/10915/26938/Documento_completo.pdf?sequence=1. Accessed 16 Oct 2013.

Svampa, M. (2012). Consenso de los commodities, giro ecoterritorial y pensamiento crítico en América Latina. *OSAL XIII, 32*.

Svampa, M., & Antonelli, M. (Eds.) (2009). *Minería transnacional, narrativas del desarrollo y resistencias sociales*. Buenos Aires: Biblos.

Teubal, M. (2009). Expansión de la soja transgénica en Argentina. In M. Pérez (Ed.), *Promesas y peligros de la liberalización del comercio agrícola: Lecciones desde América Latina* (pp. 73–91). La Paz: AIPE, GDAE.

Teubal, M. (2006). Expansión del modelo sojero en la Argentina. De la producción de alimentos a la producción de commodities. *Realidad Económica, 220*, 71–96.

Toledo López, V. (2013). Los agrocombustibles como eje del extractivismo en la Argentina. In N. Giarracca & M. Teubal (Eds.), *Actividades extractivas en expansión ¿Reprimarización de la economía argentina?* (pp. 137–158). Buenos Aires: Antropofagia.

Toledo López, V. (2016). Desarrollo y agroenergía. Un análisis de narrativas regionales y locales a propósito de la producción de biodiesel en Santiago del Estero. In G. Merlinsky (Ed.), *Cartografías del conflicto ambiental en Argentina 2*. Buenos Aires: CLACSO.

Toledo López, V. (2017). La política agraria del kirchnerismo. Entre el espejismo de la coexistencia y el predominio del agronegocio. *Mundo Agrario, 18*(37).

Toledo López, V. (2018). Agroenergía en Argentina: una discusión sobre la renovabilidad y el despojo. In F. Gutiérrez (Ed.), *Soberanía energética, propuestas y debates desde el campo popular* (pp. 117–147). Buenos Aires: Ediciones del Jinete Insomne.

UMSEF (2016). *Monitoreo de Bosque Nativo*. Buenos Aires: PEN.

Varesi, G.A. (2016). Tiempos de restauración. Balance y caracterización del gobierno de Macri en sus primeros meses. *Realidad Económica, 302*, 6–34.

Wehbe, M., Civitaresi, M., & Tarasconi, I. (2008). Promoción de agrocombustibles, oportunidades para la agricultura familiar en la provincia de Córdoba. IV Congreso Internacional de la Red SIAL, 7–31 October, Mar de Plata, Argentina.

12 Territorial Changes Around Biodiesel ... 261

Open Access This chapter is licensed under the terms of the Creative Commons Attribution 4.0 International License (http://creativecommons.org/licenses/by/4.0/), which permits use, sharing, adaptation, distribution and reproduction in any medium or format, as long as you give appropriate credit to the original author(s) and the source, provide a link to the Creative Commons license and indicate if changes were made.

The images or other third party material in this chapter are included in the chapter's Creative Commons license, unless indicated otherwise in a credit line to the material. If material is not included in the chapter's Creative Commons license and your intended use is not permitted by statutory regulation or exceeds the permitted use, you will need to obtain permission directly from the copyright holder.

Part V

The Extractive Side of the Global Biomass Sourcing

13

Contested Resources and South-South Inequalities: What Sino-Brazilian Trade Means for the "Low-Carbon" Bioeconomy

Fabricio Rodríguez

13.1 Introduction: Bioeconomy and South-South Inequalities

Shifts in energy consumption away from the early industrializing centres of the Global North towards the emerging economies of the Global South lead to key questions regarding the bioeconomy. To what extent are bioeconomy agendas shaping the transition away from fossil dependence in the context of South-South relations? Does the bioeconomy hold the potential to restructure the current landscape of global inequalities through the active engagement of actors from the Global South? This paper addresses these questions through a qualitative analysis of trade relations between the People's Republic of China (from now on China) and Brazil from 2000 to 2018. The Sino-Brazilian case is interpreted as a key axis of South-South economic exchange (Hochstetler

---

F. Rodríguez (✉)
Institute of Sociology, Friedrich Schiller University Jena, Jena, Germany
e-mail: fabricio.rodriguez@uni-jena.de

2013) with far-reaching yet largely unexplored implications for the bioeconomy. Brazil is a crucial source of biomass, metals and fossil resources for China, and is also responsible for 60% of the Amazon, an extremely important source of carbon storage. At the same time, China's increasing reliance on natural resource imports is having a remarkable impact on Brazilian efforts to promote low-carbon transitions, both domestically and internationally.

This chapter focuses on the Sino-Brazilian trade axis and its socio-ecological as well as political implications for the emergence of a global, low-carbon bioeconomy. It uses an analytical perspective that draws on insights from political geography (Andresen 2010; Bridge 2013) and studies on ecologically unequal exchange (Bunker 1990; Hornborg 1998; Frey et al. 2019). A cross-fertilizing approach to these perspectives highlights the socio-spatial and transnational dynamics of resource extraction while understanding commodity trade in terms of its political and economic relationship with society and nature.

Although issues of trade speak unsurprisingly to historical-materialistic analyses (Wallerstein 1979), this article also focuses on the struggles over meaning in which transitional agendas are embedded. This includes the "low-carbon" bioeconomy on the Brazilian side (Biofuture Platform 2018) and "ecological civilization" on the Chinese side (The State Council 2015). Hence, the analysis of trade flows draws the reader's attention to what Gavin Bridge has termed "the making of resources". This refers to "the political, economic and cultural processes through which particular configurations of socionature become imagined, appropriated and commodified" (Bridge 2011b, p. 821). As the chapter engages with the material qualities of Sino-Brazilian trade, I take an interpretive stance to the analysis of trade statistics. The analysis builds on the results of my doctoral dissertation (Rodríguez 2018), and additional research conducted in Brazil and China between 2018 and 2019.

13.2 South-South Cooperation and Energy Consumption

The wording of "South-South" cooperation suggests the emergence of a new economic and political order. In this new order, solidarity and equity—not dominance and inequality—are depicted as the principles guiding higher levels of interdependence between nations and regions of the Global South. Indeed, "South-South" discourse, vividly diffused by the BRICS (Brazil, Russia, India, China and South Africa), creates a juncture of emancipatory momentum for actors seeking to transform "the (historical) structural constraints within which they are operating" (Muhr 2016, p. 632). However, this wording freezes the meaning of South-South cooperation as an inherently "good" and "empowering" project that benefits different geographies on equal terms. While China's new prominent status in the global economy destabilizes the dominance of the West in economic globalization, China's rise cannot be equated with the rise of the Global South as a whole (Rodríguez 2018, 2020).

China's changing place in the world becomes evident when energy consumption is used as an indicator of economic activity. In 2000, for instance, the US still figured as the world's largest consumer of energy, with a total energy demand of 2269 million tonnes of oil equivalent (Mtoe) from coal, gas, oil, electricity, heat and biomass.[1] This represented almost double the amount of energy consumed in China (1161 Mtoe). Between 2000 and 2009, however, the US began to give up its position as the largest single energy consumer in the world. By 2009, China's primary energy demand had surpassed that of the US; in contrast, economic activity contracted in most Western economies because of the 2008/2009 global financial crisis. China's energy demand continued expanding until 2014 when the country stabilized its expanding dynamics in the face of the "new normal" of lower growth rates. Interestingly, the energy consumption of the BRICS surpassed that of the early industrializing nations of the G7 (Canada, France, Germany, Italy, Japan, UK and US) during the same period.

---

[1] Enerdata. Global Energy Statistical Yearbook 2016. See, https://www.enerdata.net/publications/world-energy-statistics-supply-and-demand.html. Accessed 5 June 2020.

However, as China accounts for 60% of the energy consumed by the BRICS, this acronym is far from representing a homogenous block, as illustrated in Fig. 13.1.

According to these data (Fig. 13.1), in 2018, China's energy demand was 3.4 times that of India, 4 times that of Russia, 11 times that of Brazil, and 23 times that of South Africa. In 2018, China's total energy demand was not only 1.4 times greater than that of the US, but was almost equivalent to the total energy consumed by the second and third largest energy consumers in the world (US and India, respectively). This is an important fact, given that China and India are commonly mentioned as the drivers of shifting energy geopolitics, despite the fact that the structural asymmetries between the two are considerable (Rodríguez 2018, 2020). In sum, the unqualified notion of South-South cooperation is

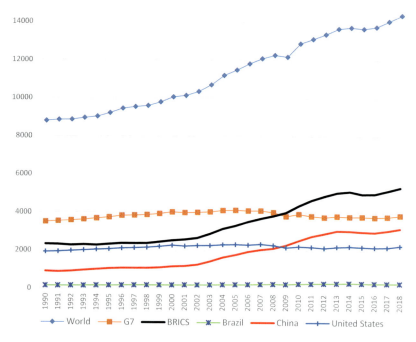

Fig. 13.1 Shifting dynamics in Global Energy Consumption, 1990–2018 [Mtoe] (*Source* Enerdata (2015, 2019); Global Energy Statistical Yearbook (2018), Accessed 1 April 2020. Author's illustration)

analytically misleading because it portrays the idea of an "equal footing" whereas, in reality, new inequalities are in fact in the making.

13.3 Going Global? Brazil Pushes for a "Low-Carbon" Bioeconomy

One of the problems of the global energy mix is its persistent reliance on carbon-intensive, non-renewable resources that are taken from the ground, combusted and carelessly emitted into the atmosphere as carbon dioxide (CO_2) (Bridge 2011a, p. 310). The project of "decarbonizing" the global economy is thus a key item in the Paris Agreement and a crucial endeavour to fight climate change. Following this rationale, the concept of the bioeconomy is part of a larger compound of concepts promising sustainable transitional pathways such as the *green* and/or *circular* economy. The bioeconomy's particularity is its focus on biomass, bioprocesses and biotechnologies as a means of substituting (part of) the petrochemical basis of current modes of production (Birch and Tyfield 2013; Backhouse et al. 2017; Goven and Pavone 2015). International organizations such as the International Energy Agency (IEA) consider Brazil a precursor of the bioeconomy because sugarcane ethanol has become one of its main energy sources for transport over the past five decades.[2] This development was a response to the fourfold increase in the price of oil. In 1975, this led Brazil's military dictatorship to launch the ethanol program *Proálcool* to reduce the country's dependence on foreign oil and enhance energy security (Wilkinson and Herrera 2010, p. 750).

Given this trajectory, Brazilian actors from industry and government ascribe their country a leading role in the global debate on bioeconomy. While Brazil has no official strategy, the Ministry of Science, Technology and Innovation (MCTIC) emphasizes the importance of biomass and bioenergy for the development of the Brazilian bioeconomy (MCTIC 2016). The National Confederation of Industry (CNI) has also issued

---

[2] IEA Renewable Energy Market Report 2018. See, https://www.youtube.com/watch?v=_8L16vS V6tM [min 14:36]. Accessed 16 Sep 2019.

a document advocating higher levels of public investment in biotechnology and biomedicine in addition to the development of a favourable regulatory framework (Harvard Business Review 2013). Due to growing international awareness about the negative ecological consequences of fossil combustion, Brazilian policymakers and agribusinesses, collaborating partly with oil transnationals, have turned to the deployment of biomass as a *green* source of energy.[3] In their view, the fight against climate change represents a new opportunity to revitalize and rescale the deployment of ethanol beyond Brazil's national borders (Lorenzen 2019, p. 8). The launching of the Biofuture Platform illustrates this recent trend. With its mission to "accelerate the transition to an advanced, low carbon, global Bioeconomy", the Brazilian government successfully promoted the creation of a multi-stakeholder platform encompassing 20 different states at the sidelines of COP 22 in Marrakech in 2016. A key item on the Biofuture agenda is the promotion of "sustainable biomass", which is said to provide a "low-carbon" solution to the material constraints of the fossil economy.[4]

However, critics have long pointed to the fact that land-based energy may be a cure that causes more harm than the actual disease (Houtart et al. 2010; Dietz et al. 2014; Holt-Giménez and Shattuck 2009; Oliveira et al. 2017). In this light, the term "agrofuels" has emerged in opposition to the term "biofuels" to highlight the fact that biofuel production hinges upon a land-intensive, largely exploitative system of monocultural plantations (Holt-Giménez and Shattuck 2009). At the same time, local people, and this also applies to those directly and/or indirectly affected by Brazil's sugarcane sector, face increasingly deteriorating conditions regarding access to land and employment (see Lorenzen in this volume).

Nonetheless, faced with the goal of ensuring that average temperatures do not rise by more than 1.5 degrees Celsius, Brazilian firms are reclaiming the relevance of sugarcane ethanol as a global alternative to

---

[3] Statements by an expert from the Brazilian Institute of Petroleum, Natural Gas and Biofuels (IBP) and a technical staff member of the National Sugarcane Association (UNICA) at the International Conference Rio Oil & Gas 2018. Energy to transform. Rio de Janeiro, 24–25 Sep 2018.

[4] See, http://biofutureplatform.org/. Accessed 3 May 2020.

fossil fuels in the transport sector (Moreno 2016). According to Biofuture, the "low-carbon" qualities of biomass stem from the assumption that the amount of CO_2 emitted through biofuel combustion is actually compensated for by the carbon sequestrating function of the plants delivering the biomass itself. However, both common sense and empirical evidence speak against the carbon neutrality of agrofuels. As such, the issue is not whether agrofuels are a source of renewable energy, but whether the global upscaling of land-based energy sources holds any realistic potential of replacing the petrochemical basis of the global economy while improving its ecological balance. Empirical research suggests that the global upscaling of crop-based fuels may increase carbon emissions, particularly if market pressure "pushes" production into new agricultural frontiers or, even worse, into the Amazonian rainforests (Gibbs et al. 2008, p. 5). The use of conventional feedstock to produce bioplastics, another aspect of Biofuture's bioeconomy concept, could have similar effects (Escobar et al. 2018, p. 11). To be clear: in Brazil, the main crop pushing deforestation in the Amazon is soy, which is mainly used by the meat industry (Trase 2018). Unfortunately, in November 2019, Jair Bolsonaro, the far-right Brazilian president, put an end to the ecological zoning of sugarcane via presidential decree. This cleared the way for sugarcane firms to expand the frontier into the Brazilian wetlands of the *Pantanal* and the Amazon (Ferrante and Fearnside 2019).

Given past and present conditions, Brazil's push for a global bioeconomy has encountered resistance both inside and outside of its borders. In an open letter, 117 civil society organizations denounced the contradictions of the "low-carbon" narrative:

> […] the BioFuture Platform advocates transitioning the energy, transport, and industrial sectors to bioenergy and biomaterials. This ignores the science – burning biomass for energy releases as many emissions as burning coal, while the production and consumption of biofuels, bioplastic or other biomaterials reduces land available for crops, leads to deforestation and other land conversions, and releases nitrous oxide.
>
> To mitigate the worst effects of climate change, we need governments, NGOs, academia, and the private sector to work together to reduce overconsumption of energy and [to] decarbonize the energy, transport, and

industrial sectors – not merely allow the rich to continue over-consuming whilst transitioning to another carbon-intensive resource".[5]

As these debates show, Brazilian support of a global, "low-carbon" bioeconomy has given way to a new discursive arena in which the renewable and hence "green" qualities of biomass are constructed in opposition to the exhaustible and polluting qualities of oil. However, there is no evidence that agrofuels can deliver a substantial contribution to the problems created by fossil fuels—certainly not on a global scale. According to Kean Birch, the three main characteristics of bioenergy consist of "low energy density, biomass conversion limits, and land footprint" (Birch 2019, p. 116). Since land intensity is a problem, there are also efforts to foster new generations of "drop-in biofuels from algae and synthetic biology" (ibid.). Notwithstanding this fact, it is highly improbable that these technologies will be able to replace the energy input provided by oil. As noted by Tiziano Gomiero (2015, p. 8491), the energy density of ethanol is simply too low for energy return on investment (EROI) to make sense in the long-run. Birch makes a similar point:

> [B]iomass couldn't possibly be used to power all sectors (e.g. heat, motor fuels, electricity) under existing rates and trends of global energy consumption – almost all estimates suggest that there just isn't enough solar energy being converted into biomass quickly enough, nor can biomass be extracted intensively enough, to allow that type of scenario to be sustainable. (Birch 2019, p. 115)

The global upscaling of biomass as a project to substitute the material basis for sustainable transport, as intended by the Brazilian government and sugarcane industry, is indeed a highly problematic agenda. At Biofuture, these actors contend that ethanol second-generation (E2G), which involves the use of microorganisms (such as algae), will help mitigate these problems. However, the future of Brazilian E2G remains uncertain due to sharply declining levels of public funding since the

---

[5] Open Letter to Biofuture. The industrialization of the Bioeconomy poses risks to the climate, the environment and people. See, https://environmentalpaper.org/wp-content/uploads/2018/11/BioFuture-Platform-Open-Letter-final-1.pdf. Accessed 5 June 2020.

2008/2009 global financial crisis, the enduring dominance of ethanol first-generation (E1G) stakeholders and the emergence of new varieties of sugarcane (Backhouse 2020, pp. 14–16). The differences between the advocates and critics of Biofuture are indeed struggles over the meaning and signification of agrofuels in terms of their "carbon neutrality". Paradoxically, the Brazilian state-led oil company Petrobrás has also questioned the government's plan to expand the use of biomass as a source of "low-carbon" energy until 2030 (Teixeira 2017). Through the RenovaBio policy, a set of regulations that enforce blending targets, Brazilian stakeholders from government, the sugarcane industry and agribusiness hope to revitalize the bio-based energy sector while claiming to tackle climate problems in the process (Backhouse 2020, p. 17). In contrast to the ethanol sector, Petrobrás executives have suggested that an increase in first-generation agrofuels production could hold Brazil back from achieving its 2030 climate objectives.[6] These encompass a 43% reduction in greenhouse gas emissions, zero deforestation in the Amazon and 45% renewables in the energy mix.

13.4 Carbon-Intensive: Sino-Brazilian Trade from a Bioeconomy Perspective

The problems and contradictions of the Brazilian government's agenda to upscale and internationalize the bioeconomy based on the explicit promotion of ethanol are further exposed by analysing Sino-Brazilian trade. If the bioeconomy is meant to promote the shift away from the deployment of fossil resources, then a much broader concept and effort to understand where to tackle this transition is required. This effort cannot only focus on the promotion of a particular resource that benefits a particular industry, such as the Brazilian sugarcane industry. Instead, a serious effort to search for an ecological balance needs to focus on the overall material base and path-dependencies linked to the fossil

---

[6] Petrobrás. *RenovaBio - Diretrizes Estratégicas para Biocombustíveis*. Relatório Técnico. See, http://www.mme.gov.br/documents/36224/930011/participacao_pdf_0.3717470965729722. pdf/8d3c6b05-80d7-372f-b1fa-596db8683174. Accessed 5 June 2020.

mode of production and identify ways to alter them. Besides, different understandings of sustainability and bioeconomy may prevail in different contexts.

For instance, Chinese officials view the concept of the bioeconomy as a Western idea with restricted potential to contribute to the "decarbonizing" of the national energy mix and to the "re-engineering" of the Chinese transport sector, in particular. The concept of the bioeconomy is not seen as providing a fitting solution to the challenges facing China. In domestic terms, the main driver is expected to be nuclear energy (NDRC—*National Development and Reform Commission* 2016), which is considered a clean source by the Intergovernmental Panel on Climate Change (IPCC), at least in terms of carbon emissions.[7] Currently, biomass makes up 2% of the Chinese energy mix. An important issue is the use of organic residuals—also referred to as "biowaste" for the production of "biogas", particularly but not only in urban areas. In this regard, the concept of the circular economy may have a higher level of relevance, whereas the bioeconomy is more likely to be developed in the fields of biochemical, biomedicine and biomaterials. The use of agrarian lands for the cultivation of energy crops is politically sensitive, since the Chinese government is likely to prioritize the use of fertile land to produce food instead of energy. Additionally, it is not practical to expect China to import South American ethanol, because the amount of fossil energy required to ship bio-based fuels across the ocean might result in a negative EROI balance.[8]

In China, a relevant concept of how to understand and, hence, tackle the current planetary crisis is the idea of "ecological civilization" (The State Council 2015). Chinese officials have begun to use this term suggesting that China may have its own way of dealing with domestic and, by extension, global ecological problems. Ecological civilization has ancient origins and reaches back 2500 years to Lao Tze. Tze depicted humans' relationship with nature as one in which the laws of society

---

[7] Discussion round at the Center for International Energy Development at Xiamen University, 9 May 2019, Xiamen, China. See also World Nuclear Association: *Nuclear Power in China.* See, https://www.world-nuclear.org/information-library/country-profiles/countries-a-f/china-nuclear-power.aspx. Accessed 6 Aug 2020.
[8] Ibid.

develop in harmony with the laws of the earth and heaven, whereas these move according to the major "Tao" (divine path), which in turn follows the course of nature (Pan 2016, p. 35). Recently, however, the use of the concept has changed and it is now much more pragmatically framed in terms of the Chinese Communist Party's (CCP) policy goals for synchronizing environmental policy and economic growth (The State Council 2015). Through its emphasis on ecological civilization, the Chinese government urges Chinese companies to accelerate and intensify efforts to build an adaptive, knowledge and technology-oriented pathway towards a low-carbon milieu for the advancement of the "green industries" (Geall and Ely 2018, p. 1187). Thus, this concept has not only framed the main narrative for environmental policy within, but also, and perhaps just as decisively, outside of China.

One example is China's Second Policy Paper on Latin America and the Caribbean (MFA—Ministry of Foreign Affairs 2016), which mentions ecological civilization as one of the main areas of cooperation without much detail. However, one thing is clear: in China, ecology has climbed up the ladder of policy priorities. Domestically, this involves a political shift away from Deng Xiaoping's "opening up" paradigm, which was based on the then much more accepted notion of "developing first and cleaning up afterwards". The new paradigm is a top-down (but also bottom-up) approach to "synchronizing growth and environmental protection".[9] This shifting reality goes hand in hand with large-scale investments in renewable energy such as wind, solar and hydropower and China's increasingly authoritarian pathway to growth and national rejuvenation under President Xi Jinping. Internationally, China has sought to fill the leadership vacuum since the Trump government decided to pull the US out of the Paris Agreement. While this situation may change with a new administration in Washington, the Chinese government is likely to continue advancing its own ecological paradigms in different multilateral arenas, including not least the Belt and Road Initiative (BRI).

[9]Discussion round at the Center for International Energy Development at Xiamen University, 9 May 2019, Xiamen, China.

The narratives of "ecological civilization" in China (The State Council 2015) and that of the "low-carbon" bioeconomy in Brazil (Biofuture Platform 2018) reflect the construction of potentially powerful discourses of political and economic change regarding nature-society relations on a global scale. Despite different framings, these two concepts address the same problem: the perceived urgency to secure current structures of growth and wealth while simultaneously reducing the carbon footprint. In 2012, Brazil and China upgraded their diplomatic relations to the level of a "global strategic partnership". Both countries opted for this format of South-South cooperation in the aftermath of Rio + 20. During the Rio summit, China and Brazil emphasized the importance of ensuring that their catching-up processes were not jeopardized by the environmental problems caused by the early industrialized nations.

In view of this situation, issues related to the bioeconomy have gained some relevance in Sino-Brazilian relations, mostly due to Brazilian pressure. The ten-year bilateral cooperation plan (2012–2021)[10] documents a bilateral commitment to the promotion of joint research and development (R&D) programmes in key areas encompassing biotechnology, bioenergy and biomedicine. On both sides, the entities holding responsibility for the implementation of these programmes consist mainly of large actors from the agrochemical industry and governmental research institutions. In this context, the question is whether Sino-Brazilian cooperation in the bioeconomy holds the potential to enhance the ecological balance of the planet while maintaining national economic growth in motion. A look at the Sino-Brazilian trade axis shows that this relationship not only resembles a new pattern of structural inequality, but it is also deeply entrenched in the prevalent structures of the fossil economy.

Given China's increasing reliance on external sources of agricultural products, minerals and oil on the one hand, and Brazil's privileged endowments in these rubrics on the other, bilateral trade has expanded enormously albeit narrowly and unevenly. Consulted trade data reveal

---

[10]The plan was signed in April 2011 by the Brazilian and Chinese Governments. For the Portuguese version, see Plano Decenal de Cooperação Brasil-China 2012–2021. In E. Moreira Lima (Ed.), *Brasil e China: 40 anos de relações diplomáticas. Analises e Documentos* (pp. 405–431). Brasília: FUNAG. http://funag.gov.br/loja/index.php?route=product/product&product_id=844. Accessed 20 July 2019.

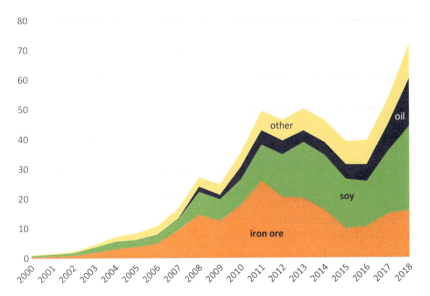

Fig. 13.2 Brazilian exports to China by commodity, 2000–2018 [Billion US$] (*Source* Chatham House (2018), 'resourcetrade.earth', http://resourcetrade.earth/. Accessed 1 April 2020. Author's illustration)

that Brazilian exports to China have expanded from US$ 1.2 billion in 2000 to US$ 63.4 billion in 2018. Similarly, Brazilian imports from China have gone from US$ 1.3 billion in 2000 to US$ 33.9 billion in 2018.[11] In terms of sheer weight, Brazilian exports to China have grown by a factor of 20 going from 17.7 million tonnes in 2000 to 346 in 2018. Trade data show that three commodities account for 82% of Brazilian exports to China (Fig. 13.2). In terms of value, soybeans accounted for 41%, crude oil for 22% and iron ore for 12% of total Brazilian exports to China in 2018.[12] Chinese exports to Brazil, in contrast, consist mainly of electronic merchandise, nuclear technologies, machinery and organic chemicals with significantly higher levels of value added, so that an unbalanced pattern of territorial specialization in the shifting configuration of the global economy is evident (Rodríguez 2018).

---

[11] The Growth Lab at Harvard University. The Atlas of Economic Complexity. See, http://www.atlas.cid.harvard.edu. Accessed 2 Nov 2020.
[12] ResourceTradeEarth. See, https://resourcetrade.earth/. Accessed 3 June 2020.

While Brazilian trade with China has contributed to the diversification of global markets away from the US—and hence reduced North-South inequalities in the economic realm—it has also meant an increasing level of trade dependency upon the Chinese market. In 2000, for example, the US market was still the main destination for Brazilian exports. Remarkably, Sino-Brazilian trade took off between 2002 and 2008, and experienced a substantial expansion between 2009 and 2013. The expanding dynamics of Brazilian trade with China were hardly affected by the 2008/2009 global financial crisis, but trade between Brazil and the US contracted considerably. As a result, China displaced the US as Brazil's most relevant export market in 2009. Sino-Brazilian trade was further consolidated between 2011 and 2015. During this period, China attracted 18.6% of Brazilian exports, whereas the US market accounted for just 12.7%. In 2018, China concentrated an overwhelming share of 27% of all Brazilian exports while the US accounted for only 11%.[13]

However, the Sino-Brazilian trade axis is constituted by China as an increasingly wealthy and politically powerful country at the core of global energy and resource consumption, and Brazil as a semi-peripheral provider of raw materials with a rapidly deteriorating status in the global economy. Moreover, from a bioeconomy perspective, the thickening flows of raw materials travelling from Brazil to China show that this axis of South-South trade is far from being "low carbon". In fact, the overwhelming predominance of soy, iron ore and oil as the bulk of Brazilian exports to China exposes three blind spots in the Brazilian push for a global bioeconomy agenda and questions the ecological balance of this important case of South-South trade inequality.

Soy

In Brazil, the expansive cultivation of soybeans is tightly linked to the rising consumption of protein by the emerging Chinese middle classes (Wilkinson et al. 2016) and China's decisive role in the global geopolitics of food (Oliveira 2015). This massive increase in biomass demand reinforces the large-scale system of industrial agriculture, which was established under colonial rule, further intensifying and aggravating the

---

[13]The Growth Lab at Harvard University. The Atlas of Economic Complexity. See, http://www.atlas.cid.harvard.edu. Accessed 3 June 2020.

land- and capital-intensive expansion of monocultures. The expanding cultivation of soy takes up 28 million hectares of land—three times more than sugar and eight times more than coffee (Wilkinson et al. 2016). Between 2013 and 2018, China's demand for soybeans from Brazil grew by 14%.[14] In order to stay profitable, these dimensions require a great degree of mechanization and digitalization, as the case of soy cultivation demonstrates. This mode of agroindustrial production has far-reaching consequences for the entire agricultural sector in Brazil. The current monoculture model leads to the massive application of pesticides and includes the genetic altering of seeds, which is a driver of soil deterioration and vast biodiversity loss. Recent studies (Trase 2018) confirm that the expansion of soy will produce dramatic changes in land use, and that this will exacerbate deforestation rates in the Amazon while pushing CO_2 emissions to critical levels. Although the territorial and commercial dynamics of sugarcane cannot be directly equated with those of soy, there are important global trade issues that demonstrate an increasing level of market-interdependence between both commodities. As China retaliates on US tariffs by cutting its imports of US-grown soy, Brazilian farmers in the sugarcane industry may adapt their production according to the developments in the soy business (Teixeira 2018).

Iron Ore

The problematic gap between the "low-carbon" narratives of the Brazilian bioeconomy agenda and the material qualities of Sino-Brazilian trade is further exposed by the flows of iron ore travelling to China's ports. Between 2013 and 2018, these grew by 7.8%.[15] Iron ore is representative of the unhindered extraction of finite resources that build the material basis for the fossil-based paradigm pertaining the American Way of Life (Backhouse et al. 2019, p. 17), and its wide-ranging effects on China's own developmental pathway: cars, urbanization, accelerated industrialization and massive consumption. Yet there is no historical precedent for the rate and scale at which imported raw materials are being processed in China. China is now responsible for more than half

---

[14] ResourceTradeEarth. See, https://resourcetrade.earth/. Accessed 3 June 2020.
[15] ResourceTradeEarth. See, https://resourcetrade.earth/. Accessed 3 June 2020.

of the global demand for minerals, which, in turn, leads to the creation of extractive enclaves in Brazil and elsewhere (Rodríguez 2018, 2020).

Oil

Of all problems, the untamed extraction of oil is probably the most pressing from a bioeconomy perspective. With 42% growth between 2013 and 2018, this source of fossil energy is adding huge amounts of carbon into the atmosphere. In addition, it is also delivering the material basis for the unequal consolidation of the Sino-Brazilian trade nexus. In 2006, Brazil discovered the second largest crude oil reserves in South America. Located at a depth of 7000 m and 300 km off the Atlantic shore of Brazil, these reserves named *Pré-Sal* (Schutte 2013) cannot possibly go unnoticed in debates about the transformational potential of the Brazilian bioeconomy. Petrobrás emerged as a global player in the crude oil business through enormous amounts of Chinese capital flowing to Brazil in the form of conditional loans and direct investments. In exchange, Brazil granted China guaranteed shipments of oil, hence easing China's energy needs (Alves 2013; Rodríguez 2018). As a result, the material basis of the Sino-Brazilian nexus remains hostage to the exact fossil structures that bioeconomy agendas are supposed to overcome (Backhouse et al. 2019).

13.5 Conclusion

This article engaged with the Brazilian project of building a global, low-carbon bioeconomy, its interconnections with the Chinese idea of ecological civilization and the making of South-South inequalities in the realm of trade. The analysis provides evidence of four important problems regarding the Brazilian government's agenda to stimulate and lead the global transition towards a global, low-carbon bioeconomy. First, the scope and focus of Brazil's international bioeconomy agenda are far too narrowly defined if a transition away from fossil fuels is to be achieved on a global scale. A narrow focus on the promotion of biomass to supply the future energy needs of the global transport sector fails its target because a systemic reduction in oil consumption is equally, if not much more urgently, needed. Second, the global upscaling of agrofuels is not an environmentally viable project. The world cannot seriously expect to solve both its energy and ecological problems by fuelling the expansion of monocultures in much-needed sink areas. An effective bioeconomy agenda should not only target the transition away from oil drilling but also reconnect agricultural practices with life-sustaining and reproductive cycles. This means embracing and rethinking the principles of a balanced relationship between society and nature as set out in ancient conceptualizations of ecological civilization. The protection of primary tropical forests and the rehabilitation of cleared areas through agroforestry systems offer just two potential examples of how this abstract idea can be translated into policy and practice. However, this potential serves no cause if the untamed logics of extraction that structure the global economy remain unchallenged; this is a pending task for current bioeconomy agendas. Third, the analysis of the Sino-Brazilian agenda for South-South cooperation reveals the predominant role of national and sector-specific interests. Whereas the Chinese government seeks to satisfy its urgent domestic requirement for oil, Brazilian agribusinesses aim to expand the international market for sugar-based ethanol and soy. Instead of facilitating Brazil's transition towards the bioeconomy, Chinese resource imports from Brazil reinforce Brazil's embeddedness in the fossil structures of the global economy. Fourth, the analysis of Sino-Brazilian trade is indicative of a new pattern of global inequality, in which Brazil's geographies of resource extraction including oil, iron ore and soy provide the material basis for China's economic growth and—by extension—the stability of its regime. In conclusion, conceptualizations of a global bioeconomy should consider why and how the political economies of distant and unequally interconnected geographies prevent the implementation of a much-needed transition to a low-carbon society.

References

Alves, A.C. (2013). Chinese Economic Statecraft: A Comparative Study of China's Oil-Backed Loans in Angola and Brazil. *Journal of Current Chinese Affairs*, *1*(42), 99–130. https://dr.ntu.edu.sg/bitstream/10356/79414/1/JCCA-2013-01-Alves-Angola.pdf%20final.pdf. Accessed 5 Nov 2020.

Andresen, M.A. (2010). Geographies of International Trade: Theory, Borders, and Regions. *Geography Compass*, *4*, 94–105.

Backhouse, M. (2020). The Knowledge-Based Bioeconomy in the Semi-Periphery: A Case Study on Second-Generation Ethanol in Brazil. *Working Paper 13, Bioeconomy & Inequalities*, Jena. https://www.bioinequalities.uni-Jena.de/sozbemedia/wp/workingpaper13.pdf. Accessed 1 June 2020.

Backhouse, M., Lorenzen, K., Lühmann, M., Puder, J., Rodríguez, F., & Tittor, A. (2017). Bioökonomie-Strategien im Vergleich: Gemeinsamkeiten, Widersprüche und Leerstellen. *Working Paper 1, Bioeconomy & Inequalities*, Jena. https://www.bioinequalities.uni-Jena.de/sozbemedia/neu/2017-09-28+workingpaper+1.pdf. Accessed 10 June 2019.

Backhouse, M., Rodríguez, F., & Tittor, A. (2019). From a Fossil Towards a Renewable Energy Regime in the Americas? Socio-Ecological Inequalities, Contradictions, and Challenges for a Global Bioeconomy. Working Paper 10, Bioeconomy & Inequalities, Jena. https://www.bioinequalities.uni-jena.de/sozbemedia/WorkingPaper10.pdf. Accessed 3 June 2020.

Biofuture Platform. (2018). Creating the Biofuture: A Report on the State of the Low Carbon Bioeconomy. http://biofutureplatform.org/resources/. Accessed 26 Octo 2020.

Birch, K. (2019). *Neoliberal Bio-Economies? The Co-Construction of Markets and Natures*. Cham: Springer International Publishing.

Birch, K., & Tyfield, D. (2013). Theorizing the Bioeconomy. *Science, Technology, & Human Values*, *38*, 299–327.

Bridge, G. (2011a). Past Peak Oil: Political Economy of Energy Crises. In R. Peet, P. Robbins & M. Watts (Eds.), *Global Political Ecology* (pp. 307–324). London: Routledge.

Bridge, G. (2011b). Resource Geographies 1. *Progress in Human Geography*, *35*, 820–834.

Bridge, G. (2013). Resource Geographies II: The Resource-State Nexus. *Progress in Human Geography*, *38*(1), 1–13.

Bunker, S.G. (1990). *Underdeveloping the Amazon. Extraction, Unequal Exchange and the Failure of the Modern State*. Chicago and London: The University of Chicago Press.

Dietz, K., Engels, B., Pye, O., & Brunnengräber, A. (2014). *The Political Ecology of Agrofuels*. Routledge ISS Studies in Rural Livelihoods, 13. Abingdon: Routledge.
Escobar, N., Haddad, S., Börner, J., & Britz, W. (2018). Land Use Mediated GHG Emissions and Spillovers from Increased Consumption of Bioplastics. *Environmental Research Letters*.
Ferrante, L., & Fearnside, P. (2019). Sugarcane Threatens Amazon Forest and World Climate; Brazilian Ethanol Is Not Clean (Commentary). *Mongabay*. https://news.mongabay.com/2019/11/sugarcane-threatens-amazon-forest-and-world-climate-brazilian-ethanol-is-not-clean-commentary/. Accessed 3 June 2020.
Frey, R.S., Gellert, P.K., & Dahms, H.F. (2019). *Ecologically Unequal Exchange. Environmental Injustice in Comparative and Historical Perspective*. Cham: Palgrave Macmillan.
Geall, S., & Ely, A. (2018). Narratives and Pathways Towards an Ecological Civilization in Contemporary China. *The China Quarterly, 236*, 1175–1196.
Gibbs, H.K., Johnston, M., Foley, J.A., Holloway, T., Monfreda, C., Ramankutty, N., et al. (2008). Carbon Payback Times for Crop-Based Biofuel Expansion in the Tropics: The Effects of Changing Yield and Technology. *Environmental Research Letters, 3*, 1–10.
Gomiero, T. (2015). Are Biofuels an Effective and Viable Energy Strategy for Industrialized Societies? A Reasoned Overview of Potentials and Limits. *Sustainability, 7*, 8491–8521.
Goven, J., & Pavone, V. (2015). The Bioeconomy as Political Project: A Polanyian Analysis. *Science, Technology & Human Values, 40*, 302–337.
Harvard Business Review (2013). Bioeconomy. An Agenda for Brazil. http://arquivos.portaldaindustria.com.br/app/conteudo_24/2013/10/18/411/20131018135824537392u.pdf. Accessed 26 Oct 2020.
Hochstetler, K. (2013). South-South Trade and the Environment. A Brazilian Case Study. *Global Environmental Politics, 13*(1), 30–48.
Holt-Giménez, E., & Shattuck, A. (2009). The Agrofuels Transition. *Bulletin of Science, Technology & Society, 29*, 180–188.
Hornborg, A. (1998). Towards an Ecological Theory of Unequal Exchange: Articulating World System Theory and Ecological Economics. *Ecological Economics, 25*(1), 127–136.
Houtart, F., Bawtree, V., Bello, W.F., Bukassa, B.L., Geuens, G., & Mpozi, B.B. (2010). *Agrofuels: Big Profits, Ruined Lives and Ecological Destruction* (Transnational Institute Series). London: Pluto Press.

Lorenzen, K. (2019). Sugarcane Industry Expansion and Changing Land and Labor Relations in Brazil. The Case of Mato Grosso do Sul 2000–2016. Working Paper 9, Bioeconomy & Inequalities, Jena. https://www.bioinequalities.uni-jena.de/sozbemedia/WorkingPaper9.pdf. Accessed 20 May 2020.
MCTIC (2016). Estratégia Nacional de Ciência, Tecnologia e Inovação 2016–2022. Ministério da Ciência, Tecnologia, Inovações e Comunicações. Brasília. http://www.finep.gov.br/images/a-finep/Politica/16_03_2018_E strategia_Nacional_de_Ciencia_Tecnologia_e_Inovacao_2016_2022.pdf. Accessed 20 May 2020.
MFA (2016). China's Policy Paper on Latin America and the Caribbean. http://www.fmprc.gov.cn/mfa_eng/wjdt_665385/2649_665393/t1418254.shtml. Accessed 19 Sep 2019.
Moreno, C. (2016). *Landscaping a Biofuture in Latin America.* Berlin: FDCL.
Muhr, T. (2016). Beyond 'BRICS': Ten Theses on South-South Cooperation in the Twenty-First Century. *Third World Quarterly, 37,* 630–648.
NDRC (2016). The 13th Five-Year Plan for Economic and Social Development of The People's Republic of China (2016–2020). https://en.ndrc.gov.cn/policyrelease_8233/201612/P020191101482242850325.pdf. Accessed 6 Aug 2020.
Oliveira, G. de L.T. (2015). The Geopolitics of Brazilian Soybeans. *The Journal of Peasant Studies, 43,* 348–372.
Oliveira, G. de L.T., McKay, B., & Plank, C. (2017). How Biofuel Policies Backfire: Misguided Goals, Inefficient Mechanisms, and Political-Ecological Blind Spots. *Energy Policy, 108,* 765–775.
Pan, J. (2016). *China's Environmental Governing and Ecological Civilization.* Berlin, Heidelberg: Springer.
Rodríguez, F. (2018). *Oil, Minerals and Power: The Political Economy of China's Quest for Resources in Brazil and Peru.* Doctoral Thesis. University of Freiburg.
Rodríguez, F. (2020). Endstation China? Die globalen Stoffströme auf ihrem Weg durch die „Werkstatt der Welt". *PROKLA. Zeitschrift für kritische Sozialwissenschaft, 198,* 89–108.
Schutte, G.R. (2013). Brazil: New Developmentalism and the Management of Offshore Oil Wealth. *ERLACS, 95,* 49–70.
Teixeira, M. (2017, April 19). Petrobras Opposes Brazil Plan to Boost Biofuels After Selloff. *Reuters.* http://www.reuters.com/article/us-petrobras-climatechange-idUSKBN17K2MG. Accessed 12 Oct 2020.
Teixeira, M. (2018, Aug 14). Brazil's Farmers Dump Sugar for Soy as Trade War Boosts Chinese Demand. *Reuters.* https://www.reuters.com/article/us-

brazil-grains-sugar/brazils-farmers-dump-sugar-for-soy-as-trade-war-boosts-chinese-demand-idUSKBN1KZ0B5. Accessed 29 May 2019.

The State Council. (2015). Central Document No. 12: Opinions of the Central Committee of the Communist Party of China and the State Council on Further Promoting the Development of Ecological Civilization. http://www.gov.cn/xinwen/2015-05/05/content_2857363.htm. Accessed 6 Aug 2020.

Trase. (2018). Trase Yearbook 2018: Sustainability in Forest-Risk Supply Chains: Spotlight on Brazilian Soy. https://yearbook2018.trase.earth/. Accessed 20 May 2020.

Wallerstein, I.M. (1979). *The Capitalist World-Economy: Essays*. Cambridge, New York: Cambridge University Press.

Wilkinson, J., & Herrera, S. (2010). Biofuels in Brazil: Debates and Impacts. *The Journal of Peasant Studies, 37*, 749–768.

Wilkinson, J., Wesz Junior, V.J., & Lopane, A.R.M. (2016). Brazil and China: The Agribusiness Connection in the Southern Cone Context. *Third World Thematics: A TWQ Journal, 1*, 726–745.

Open Access This chapter is licensed under the terms of the Creative Commons Attribution 4.0 International License (http://creativecommons.org/licenses/by/4.0/), which permits use, sharing, adaptation, distribution and reproduction in any medium or format, as long as you give appropriate credit to the original author(s) and the source, provide a link to the Creative Commons license and indicate if changes were made.

The images or other third party material in this chapter are included in the chapter's Creative Commons license, unless indicated otherwise in a credit line to the material. If material is not included in the chapter's Creative Commons license and your intended use is not permitted by statutory regulation or exceeds the permitted use, you will need to obtain permission directly from the copyright holder.

14

Sustaining the European Bioeconomy: The Material Base and Extractive Relations of a Bio-Based EU-Economy

Malte Lühmann

14.1 European Bioeconomy—Global Biomass Sourcing?

The European Union defines a bioeconomy in its eponymous strategy as an economy based on the production and conversion of biomass: "The bioeconomy […] encompasses the production of renewable biological resources and the conversion of these resources and waste streams into value added products, such as food, feed, bio-based products and bioenergy" (European Commission 2012, p. 10). A growing bioeconomy is meant to replace a range of products currently produced from fossil resources with more sustainable alternatives based on renewable biomass. In a long-term vision, it is not only traditional goods like food and feed or paper and furniture that would be made from biogenic sources, but also industrial products ranging from chemicals and plastics to construction materials and energy.

M. Lühmann (✉)
Friedrich Schiller University Jena, Jena, Germany
e-mail: malte.luehmann@uni-jena.de

© The Author(s) 2021
M. Backhouse et al. (eds.), *Bioeconomy and Global Inequalities*,
https://doi.org/10.1007/978-3-030-68944-5_14

If this vision ever becomes a reality, the EU-economy is likely to consume more biomass than it already does. While there is no consensus in the literature about the exact scale of the demand for biomass in a future European bioeconomy, increases are expected to be significant (Scarlat et al. 2015, pp. 26–27). One study conducted on behalf of the EU, which has become a basis for official projections, found that demand for biomass can be expected to grow on a global scale by between 49% (a modest bioeconomy) to 96% (a bioeconomy boom) by 2050 (Kovacs 2015, p. 89). Only a few tentative projections exist on how to meet the increased demand for biomass. The EU Commission acknowledges this lack of information in its updated bioeconomy strategy: "Notably, information is still scarce on how much biomass is available and can be mobilized sustainably, how much is being used and for which purposes, and how the increased pressure on natural resources can be reconciled with environmental, economic and social sustainability in Europe and globally" (European Commission 2018, p. 32).

It is no coincidence that the EU is concerned about the sustainability of increased resource use on a global scale. In addition to environmental concerns, this concern is due to the fact that the EU will hardly be able to fully satisfy increasing demand for biomass without imports. The United Nations Environment Programme (UNEP) stresses this point in a study on trade flows in biomass and other resources: "At the global level, Asian and European countries are close to maximum productivity for their available land. Intensification is at a maximum and does not leave much space for further increases in productivity. These densely populated areas are depending on imports from other regions, i.e. regions of low population density" (UNEP 2015, p. 69). Along with developments in Asia, the expansion of an EU bioeconomy will thus have a significant impact not only on biomass production in Europe but also on global biomass demand and trade flows. The EU is already among the biggest biomass consumers in the capitalist world system. Growing demand for biomass imports means that other countries or regions need to export more if the bioeconomy is to flourish in Europe. These interrelations have far-reaching implications for the economic models of biomass-exporting countries, an aspect widely neglected in European bioeconomy debates. Engaging this blind spot leads to questions about the global social and environmental sustainability of expanding the bioeconomy in Europe (Ramcilovic-Suominen and Pülzl 2018, p. 4178).

Against this background, and in order to assess the consequences of a growing bioeconomy, the following analysis is aimed at answering three questions: What forms the material base of the existing EU bioeconomy? Which trade flows are relevant to the EU bioeconomy and how does the EU bioeconomy connect to economies in other countries or regions? What kind of change in these relations can be anticipated based on expectations about future biomass demand in the EU? The first question can be answered primarily in the framework of material flow accounts (MFA). The MFA system provides quantifications of biomass and other material inputs and flows that constitute the material "metabolism" of an economy (Eurostat 2018, p. 12). In contrast to conventional trade data, which commonly measures trade flows in monetary terms, MFA data highlight the material footprint of traded commodities. The second and third questions about the material connections between the EU bioeconomy and national economies around the world and about possible change to these relations calls for a transnational perspective. This is achieved using the framework provided by world systems theory (WST) (Wallerstein 2007). WST analyses the capitalist world system from the vantage point of the transnational division of labour among national economies. Wallerstein uses what he calls the "axial division of labour" to distinguish between the centre, periphery and semi-periphery of the world system (Wallerstein 2007, pp. 28–29). These concepts appear to be useful for the analysis of transnational relations in the bioeconomy, because they highlight the connections between politico-economic developments in different parts of the world. Finally, the economic dynamics of biomass-exporting countries are discussed based on debates about (neo-)extractivism as an economic and/or development model that is mainly implemented in the peripheries of the capitalist world system (Gudynas 2011; Svampa 2012; Schaffartzik and Pichler 2017).

Accordingly, the analysis is structured as follows: the second chapter begins with a brief outline of relevant theoretical concepts and their application in the context of this text. The third chapter discusses the state of the EU bioeconomy in terms of its resource usage and transnational linkages. Projections about the future biomass demand of a growing bioeconomy are presented in the fourth chapter along with an analysis of likely consequences. The fifth chapter draws conclusions and points to critical questions about transnational relations, which the future bioeconomy, as envisioned in the EU strategy, will face. Findings are based on original MFA data supplemented by other datasets, secondary analyses and further studies on resource use and biomass trade.

14.2 The Capitalist World System, Extractivism and Extractive Relations

Two key insights from WST are important for this analysis of the European bioeconomy: first, this perspective draws attention to the fact that production in capitalist economies tends to be transnational in scope and that markets tend to be connected beyond nation-state borders (Wallerstein 2007, pp. 24–27). Secondly, at the same time, the fragmentation into national economies helps to establish and protect differences in profitability and thus advantages in the accumulation process for certain capitalist enterprises (ibid., pp. 27–30). This leads national economies to become the site of more or less profitable production processes or of differently profitable steps in a given production process. Without going more into detail at this point, the global division of labour, in which more or less profitable steps of a given production process are allocated to certain national economies, constitutes centre and periphery positions in the world system. While the more profitable activities of certain production processes concentrate in the centre, less profitable activities usually take place on the periphery. The centre-periphery relation is not fixed (ibid., pp. 29–30). Changes in the composition of production processes in a national economy may lead to changes in its relative position vis-à-vis other economies in the world system. So-called *emerging economies* like the Asian Tigers (Hong Kong, Singapore, South Korea, Taiwan) or the BRICS states (Brazil, Russia, India, China, South Africa) are cases in point.

The WST perspective is helpful for the study of inequalities in the bioeconomy, because it conveys an understanding of asymmetries in transnational economic and political relations. The bioeconomy project is mainly concerned with developing "simple" primary industries into advanced production processes. The transnational (re-)distribution of more profitable steps in new bio-based value chains among production sites is a contested part of this development. Strengthening competitiveness in the global economy is a key issue of the EU bioeconomy strategy (European Commission 2018, p. 10). The focus on competitiveness can be seen as a means to defend and amplify the position of European countries at the centre of the capitalist world system. At the very least, the EU member states in Western and Southern Europe are clearly part of the core, even if the configuration of the world system itself has been changing with the relative decline of US hegemony and the ascent of China and others (Babones 2005, p. 51; Komlosy 2016, pp. 465–467).

In contrast to the EU, countries like Malaysia, Brazil and Argentina promote the bioeconomy with the explicit aim of upgrading their primary industries in order to incorporate more sophisticated steps in respective production processes (Backhouse et al. 2017, pp. 17–20). These countries' governments frame the bioeconomy as a possible means of escaping their semi-/peripheral role as subordinate suppliers of primary products for the world market. In the debate about political economy, and especially in Latin America, this role has been referred to increasingly as "extractivism" or, in conjunction with the rise of left-wing governments in the 2000s, as "neo-extractivism" (Gudynas 2011; Svampa 2012). Gudynas defines extractivism as the extraction of large amounts of raw materials in enclave economies primarily for export and with little or no domestic processing (Gudynas 2011, p. 70). Although the term is more widely used for mining and oil drilling, it is also applicable to the export-oriented agro-industrial production of biomass. Locating the debate in a world systems perspective, Svampa characterizes extractivism as a form of territorial and global division of labour between centre and periphery where Latin American countries, among others, are condemned to provide raw materials (Svampa 2012, p. 14).

Assuming a broader view that includes other world regions, Schaffartzik and Pichler focus on the transnational dimension of extractive economies to develop a quantitative analysis of extractivism based on material flows (Schaffartzik and Pichler 2017). In general, they underline the point that "[b]y supplying energy- and material-intensive resources to the global market, extractive economies enable other countries to concentrate on the addition of value in the secondary or even the tertiary sectors" (ibid., p. 1). Beyond the identification of extractive economies, the sole analysis of material flow data seems to bear few insights, as these countries are found to be quite diverse in terms of socio-economic structures and wealth (ibid., pp. 6–7). The authors emphasize that further qualitative factors need to be considered in order to assess the circumstances of extractive activities. For example, the economic structures associated with the biomass production process (e.g. monocultures, the focus on cash-crops and agro-industrial production methods) are decisive for any analysis of extractivism. Citing the example of Canada, Pichler and Schaffartzik also point to the fact that extractive economies might exist on a subnational level based on disparities between regions inside a national economy (ibid., p. 3).

Disregarding such disparities in the internal structures of extractive economies for the moment, it is still possible to highlight some points about the role of these economies in the world system and in relation to non-extractive economies like the EU. First, extractive economies produce raw materials like biomass primarily for export to the world market. Second, the consumption of these raw materials allows importing countries to concentrate on generally less energy- and material-intensive processing. Third, the allocation of the steps associated with extracting and processing in bio-based production processes to different countries constitutes a centre-periphery relation between the involved economies or subnational regions. Therefore, the relation between the EU and a biomass-supplying extractive economy can be referred to as an *extractive relation*. This relation is a centre-periphery relation based on raw material flows from primary producing countries or subnational regions to Europe. The prospects of a growing bioeconomy need to be evaluated against this background.

14.3 Biomass Flows and the EU-Economy Today

The construct of bioeconomy unites diverse economic activities from fisheries to the production of biofuels. The common denominator that defines these activities as part of a bioeconomy (at least in the European conception) is the production and conversion of biomass. Following this definition, 9% of GDP in the EU was generated by the bioeconomy in 2016 (Ronzon et al. 2017, p. 6). Biomass consumption in the EU has seen some fluctuation with a tendency towards moderate growth over the period from 2008 to 2016 for which statistical data is available (see Fig. 14.1).[1] By the end of this period, almost 2 billion tonnes (t) of biomass were being used per year (including for exports) distributed over three main categories, in addition to a neglectable share of fishery and hunting products. The bulk of biomass inputs comprises one-fifth wood, two-fifths crops (excluding fodder crops) and two-fifths crop residues, fodder crops and grazed biomass. Biogenic raw materials are used mainly for feed and food (61.93%), bioenergy (19.13%) and as biomaterials (18.82%) (Camia et al. 2018, p. 83).[2]

Overall, the share of imports accounts for 16% of total raw material inputs (RMI) in biomass with higher shares for crops and wood (21% each) and a lower share for fodder crops, residues and grazed biomass (6%) (see Fig. 14.2). Compared to other raw materials like metal ores and fossil energy carriers, import-dependency is relatively low in the biomass sector (Eurostat 2018, p. 106). However, an import contribution of 16% is still a significant amount. When the balance between imports and exports is considered, the EU has a physical trade deficit.

---

[1] Aggregate data on biomass consumption and flows is taken from the EU's material flows and resource productivity database. See, https://ec.europa.eu/eurostat/web/environment/material-flows-and-resource-productivity. Accessed 10 Nov 2020. The year 2016 is chosen as a reference for the data presented in this paper because complete data for later years is not available in all cases.

[2] Biogas and bioelectricity are not included in these figures.

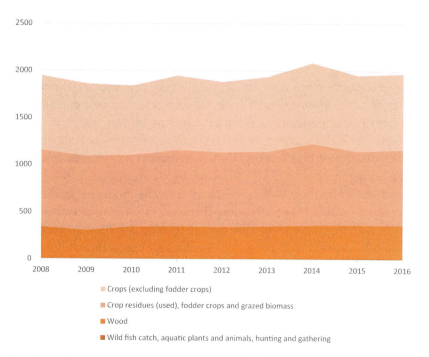

Fig. 14.1 Biomass inputs in the EU-28 over time (RMI in Million Tonnes of Raw Material Equivalent) (*Source* Eurostat)

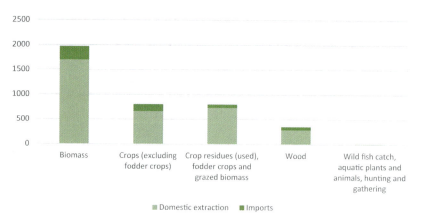

Fig. 14.2 Biomass inputs in the EU-28 by type, 2016 (in Million Tonnes of Raw Material Equivalent) (*Source* Eurostat)

Nevertheless, this deficit has shrunk since its last peak in 2007. However, imports have continued to grow but their growth was outpaced by the growth in biomass exports. In 2016, EU biomass imports were 1.22 times higher than exports (in 2007 the ratio was 1.74 to 1). Together with Asia and Africa, Europe forms the group of global net importers of biomass, whereas North America, Latin America and Oceania are net exporters (UNEP 2015, p. 60). Even though biomass is primarily consumed locally and is not as widely traded transnationally as other commodities, the biomass trade is an important and, in absolute terms, a growing segment of global markets (UNEP 2015, p. 59). Europeans are important actors in this sector as exporters and more importantly as importers of biomass. This role will be further augmented by a growing bioeconomy.

To assess the implications of European biomass use and imports in the world system context, it is important to scrutinize relations with individual countries based on these flows. Unfortunately, trade flows between the EU and individual trading partners cannot be analysed at the level of abstraction presented so far because a functional accounting system for incorporated biomass flows on a global scale that treats the EU as a block does not (yet) exist. Individual trade flows only become visible at the level of single commodities, which is why the composition of imports deserves more detailed attention. In order to identify the commodities that constitute the main biomass flows into the EU, MFA data can be disaggregated to some extent beyond the main categories. These sub-categories can then be cross-matched with data on the commodity trade from the UN Comtrade database.[3] Looking at crops (excluding fodder crops), oil-bearing crops are the biggest sub-category for imports. Among all imported oil-bearing crops in the Comtrade database, palm oil and its residues make up the largest proportion of EU-imports. For crop residues (used), fodder crops and grazed biomass, the biggest sub-category in

[3]UN Comtrade database. See, https://comtrade.un.org/data. Accessed 15 May 2019. Direct comparisons between European MFA data and UN data using HS codes are not possible because of differences in methodology, categories and units of measurement. However, absolute quantities of traded commodities are less important here. Rather, the intention is to identify the relative importance of traded commodities and especially the trade relations with individual countries constituted by these flows.

terms of imports are fodder crops and grazed biomass. Comtrade shows that soya beans including soya oil cake is by far the most important fodder crop imported to the EU. Wood places timber (industrial roundwood) as the most imported sub-category. Data from Comtrade show a slightly different picture due to a different system of categorization. In this case, the three largest imports are fuel woods, wood in the rough and wood sawn or chipped. These items are combined for the purpose of this analysis.

These considerations enable the origins of the most important biomass commodities that are imported into the EU to be identified (see Fig. 14.3). A pattern of three main biomass flows emerges from the data on individual commodities: the first flow consists of soya beans and soya oil cake, which are primarily imported from the Americas, namely from Brazil, Argentina, the USA and Paraguay. The second flow includes palm oil and its solid residues, with Indonesia and Malaysia as the two dominant importers, both of which are located in Southeast Asia. The third flow is more diverse. It encompasses imports of wood

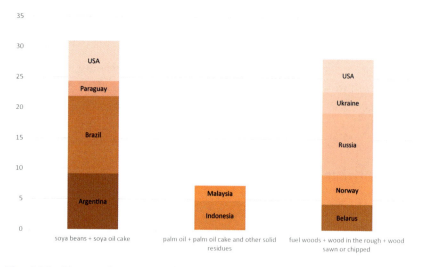

Fig. 14.3 Biomass imports to the EU-28 by country, 2016 (in Million Tonnes; Netweight for Wood) (*Source* UN Comtrade database; columns only show the biggest importers of each commodity with a combined proportion of at least 90% of imports for the respective commodity)

used for industrial and construction purposes and as an energy source. This wood is imported from the EU's northern and eastern neighbours: Russia, Norway, Belarus and Ukraine, as well as from the USA (mainly fuel woods). The three flows described here do not cover all of the biomass flows to the EU. Nevertheless, they illustrate the extractive relations between the EU and its most important trading partners in terms of biomass commodities.

Analysing the EU's external land consumption enables an assessment to be made of the importance of imported biomass for the EU and, hence, the significance of its extractive relations with its main biomass suppliers. In general, the EU is a net importer of embodied land, meaning that its consumption including exports exceeds the land-based production the EU can provide within its borders (UNEP 2015, p. 66). Quantifying the EU's external land consumption in more detail involves complex modelling, and this leads different indicators to be used. O'Brien et al. (2015) specify domestically available agricultural land as well as the land area embodied in imports and exports in absolute terms. They found that agricultural land in the EU-27 covered roughly 187 million hectares (ha) in 2011 with a slight decrease since 2000 (ibid., p. 240). A further 45 million ha of agricultural land abroad was required for imports in the same year, showing considerable fluctuation over time but with a small decrease compared to the year 2000 (ibid., p. 241). Around 19 million ha of embodied agricultural land were exported in 2011 and this trend has been increasing since 2000 (ibid., p. 241). These figures clearly show the position of the EU as a net importer but also the considerable magnitude of imported agricultural land compared to the domestically available area.

To put these figures into a global perspective, European land consumption can be examined in proportion to the globally available land area. In their study of global land use for the domestic consumption of biomass, Bringezu et al. (2012) looked at per capita consumption of croplands in Europe and around the world. They calculated that the EU-27 required 0.31 ha per capita of global cropland in 2007 although only an average of 0.24 ha per capita was available globally (ibid., pp. 227–228). The average European citizen thus consumed roughly 30% more cropland per capita than the average global supply.

For a future European bioeconomy, non-food applications of biomass are particularly important, as the widespread use of novel bio-based products in areas such as chemicals or plastics is at the heart of the project. Bruckner et al. (2019) quantified the cropland footprint of the EU's bioeconomy excluding the food sector.[4] The study concluded that the EU is even more reliant on biomass imports for non-food applications compared to overall biomass imports. As stated above, imports account for 16% of total European raw material inputs in biomass. For the non-food sector, only 35% of products were produced from domestic land resources in 2010, whereas the vast majority—65%—was based on imported croplands (ibid., p. 5). The cropland footprint of vegetable oils and oil crops represented the biggest share of imports of non-food products and make up one-third of total imports (ibid., pp. 5–6). This mostly reflects the material flows of palm oil and derived products and the extractive relations with Southeast Asian countries.

By looking into material flows, trade data and analyses of the EU's land footprint, a clear picture emerges of the external dimension of the current European bioeconomy. Today the EU-economy relies on biomass imports to provide raw materials for a variety of uses. Three major biomass flows can be identified that constitute extractive relations with peripheral and semi-peripheral countries like Indonesia and Paraguay but also with extractive activities in the USA. The EU's land footprint adds a notion of the considerable overconsumption of global agricultural land by the EU and the central importance of imports for the core area of a developing bioeconomy: the non-food sector.

14.4 Projections for a European Bioeconomy

There are many uncertainties about the biomass consumption of a future bioeconomy on the European as well as on the global level. The question as to whether enough biomass can be produced sustainably as a substitute for the current use of fossil raw materials is a highly contentious

---

[4] By taking croplands as the basic analytical category, Bruckner et al. exclude the forestry sector from their analysis.

topic (Priefer et al. 2017, pp. 7–8). The EU has been looking into the issue and commissioned its Standing Committee of Agricultural Research (SCAR) to report on biomass flows in a future bioeconomy in 2015 (Kovacs 2015). The SCAR-experts describe three scenarios for worldwide biomass demand and supply in the year 2050 (ibid., pp. 88–91). While food and feed demand are expected to remain the same over all three scenarios based on forecasts of the Food and Agriculture Organization of the United Nations (FAO), figures for biomass supply and demand for material and energy uses vary. All scenarios are compared to a 2011 baseline with a supply and demand of 12.18 billion t of dry matter of biomass. The first scenario—"Bio-Modesty"—predicts a moderate growth of biomass demand to 18.2 billion t, which is matched by the same level of growth in supply. The second scenario—"Bio-Boom"—forecasts a high level of growth in worldwide biomass supply and demand to 23.9 billion t, respectively. However, the third scenario—"Bio-Scarcity"—expects a gap to occur as a projected high demand for biomass at 23.9 billion t cannot be matched with an almost stagnant level of supply of 13 billion t. The likelihood of each of these scenarios depends, on the one hand, on uncertainties about growth in demand arising from population and economic growth, the relative scarcity of classical resources and the evolution of bio-based and non-biomass-based technologies. On the other hand, supply growth is expected to be influenced mainly by the development and implementation of new technologies (ibid., p. 88).

Overall, the scenarios underline the perceived necessity to intensify biomass production significantly in order to meet the demands of a future bioeconomy. The report also points to some broader socio-economic consequences and serious environmental risks associated with such developments (ibid., pp. 90–91). However, the report does not draw any general consequences in terms of the feasibility of growing the bioeconomy. Furthermore, while the SCAR-foresight exercise serves as a basis for EU-policy projections and is cited in the European bioeconomy strategy, the scenarios themselves assume a global view without explicitly situating the EU in the global picture. As such, it does not deal with the question of how the growth of a European bioeconomy will fit in with global developments.

In another assessment of the role of biomass in a future bioeconomy, researchers from the Commission's Joint Research Centre (JRC) come to more detailed conclusions about the EU's relative position in terms of global developments: the EU will depend on biomass imports "to provide biomass feedstock for the bio-based economy in the future. Imports will mainly consist of crops, vegetable oils, wood and wood products, wood pellets or biofuels. An increase in the bio-based economy is expected to be a worldwide development. Therefore, only a part of the globally available biomass potential is available for the EU" (Scarlat et al. 2015, p. 27). The commodities mentioned here are the same that already make up most of the EU's biomass imports today (see above).

Combining the conclusions of both the experts from SCAR and the researchers from the JRC, it seems clear that biomass demand will grow globally as well as in the EU. The EU will be dependent on imports, mainly of the same commodities that it imports today. Increased European and global demand will only be satisfied in a scenario of intensified biomass production. The EU's situation is further aggravated by the fact that eight out of the ten countries identified above as the EU's main biomass suppliers have formulated their own bioeconomy strategies or biotechnology-focused development strategies (German Bioeconomy Council 2015, pp. 126–132; German Bioeconomy Council 2018, p. 13).[5] If these strategies were to be implemented successfully, domestic consumption and processing of biomass in these countries can be expected to increase in the future. As such, exports of raw materials to the EU are likely to face increased competitive pressure from domestic processing undertaken in these countries. Concerning the transnational relations of the European bioeconomy, the questions arise as to where the additional biomass will be produced and what impact this might have on the three flows of EU imports described above.

A tentative answer can be found in a study by Piotrowski et al. (2015b), which also served as a basis for the projections provided by SCAR. As part of scenario development, Piotrowski et al. discuss the potential for expanding the land used for cultivation for biomass production. Summarizing their findings, they present three supply scenarios

---

[5] Exceptions are Belarus and Ukraine.

for the year 2050 (Piotrowski et al. 2015a, pp. 4–5). In a low supply scenario, no expansion of agricultural land is necessary. Instead, they assume a loss of 100 million ha of agricultural land due to soil degradation. This scenario corresponds to the "Bio-Scarcity" scenario cited above and does not provide enough biomass for even modest growth of the bioeconomy. In a business-as-usual supply scenario, new agricultural areas of 435 million ha are forecast based on the conversion of areas currently not used for crop production, mainly pastures and meadows but also forests. This corresponds to the "Bio-Modesty" scenario. Finally, a high supply scenario would require 760 million ha of new agricultural land by converting formerly unused land or land that was used for other purposes. Given that all the cropland available today amounts to 1,400 million ha, and that FAO data estimate total agricultural land including pastures to cover 5,000 million ha,[6] these increases would be substantial. Again, Piotrowski et al. found that the geographical distribution of potentially available areas for cultivation is uneven. Sub-Saharan Africa and Latin America are particularly mentioned as regions with a potential for increased rainfed agriculture (Piotrowski et al. 2015b, p. 111).

In summary, no definite projections about the biomass demand of a future bioeconomy in Europe or worldwide are currently available. Various scenarios can be drawn on depending on the magnitude of growth in the bioeconomy. As long as growth in the bioeconomy is expected, all of these scenarios include substantial increases in biomass production and agricultural land use. In this context, the EU will be even more dependent on imports, as the potential for domestic intensification of agriculture has mostly been exhausted (UNEP 2015, p. 69). If sub-Saharan Africa and Latin America are to intensify biomass production in the future, as projected, some material flows to the EU are likely to change, as material flows from Southeast Asia and the EU's northern and eastern neighbourhood would lose importance in relative terms. However, extractive relations with the peripheries and semi-peripheries of the world system will continue to play an important role for the European bioeconomy under these circumstances. Moreover, the development of bioeconomies in the countries that are predicted to supply

---

[6]See, http://www.fao.org/faostat/en/. Accessed 14 May 2019.

the EU with biomass questions the feasibility of an import-reliant future European bioeconomy.

14.5 Questioning the Transnational Sustainability in the European Bioeconomy

As this study has shown, the extractive relations embedded within biomass flows to the EU play a significant role in the existing European bioeconomy. The European economy relies heavily on biomass imports and will do so even more in the future especially when the resource needs of a growing bioeconomy are considered. The EU's extractive relations constitute unequal transnational linkages to those countries or subnational regions that extract raw materials for processing by European "bio-industries". These relations are hardly discussed in European bioeconomy politics. In fact, the updated European bioeconomy strategy, the most important document informing bioeconomy policy in the EU, pays very little attention to the EU's global role as an importer of biomass. Only a descriptive section on the European food-system mentions imports of several food commodities such as palm oil (European Commission 2018, p. 47). More comprehensive passages on biomass flows fail to mention the role of imports altogether (ibid., pp. 35–38). The global context is discussed primarily in terms of the challenges posed by expanding food demand outside of the OECD-world and by escalating competition over resources (ibid., p. 33). Based on this assessment, managing biomass supply is one of the five objectives drawn up by the updated strategy: "Managing natural resources sustainably, is central for a bioeconomy whose parts are increasingly interlinked. More than ever, a circular bioeconomy depends on an efficient and sustainable use of biological resources, against the backdrop of an increasing demand for biomass" (ibid., p. 26). Developing the bioeconomy in Europe is then addressed as a remedy against increasing demand: "A sustainable bioeconomy is essential to tackle climate change and land and ecosystem degradation. It will address the growing demand for food, feed, energy, materials and products due to an increasing world population, and reduce our dependence on non-renewable resources" (ibid., p. 15). While the strategy mentions the expectations of increasing pressure and competition over biomass supplies on a global scale as a concern, it fails to address the role of the EU and its bioeconomy policy as a possible driver behind these phenomena.

This blind spot leads to questions about the much-acclaimed sustainability of European bioeconomy politics, which is also expressed in the quotes above. Ramcilovic-Suominen and Pülzl (2018) resume that the EU's approach to sustainability is too narrow and that it fails to address the social aspects of sustainability: "The current EU bioeconomy policy leans strongly towards conservationist, utilitarian and instrumental approaches to SD [sustainable development], as well as to weak sustainability". Furthermore, the authors suggest that the "EU bioeconomy policy debate is likely to change the contemporary predominant policy discourse on SD by strongly emphasising technological solutions and [the principles of] economic efficiency". They point out that these solutions are likely to include new materials and products as well as a focus on competitiveness, but that they will fail "to emphasise the environmental (biodiversity, ecosystem services) and social aspects (justice, equality, benefit sharing) of SD" (ibid., p. 4178).

Tackling unequal transnational relations between centre and periphery in terms of material flows is among the social aspects of sustainability that the EU has neglected so far. The extractive relations with other world regions are likely to grow even deeper as long as the problem of insufficient global biomass supplies for a growing bioeconomy is discussed in the framework of competition. In the light of bioeconomy-based growth strategies in the Global South, such forms of transnational extractivism appear particularly problematic. Therefore, instead of increasing the use of biomass in order to substitute fossil-based raw materials, Europe needs to reduce its consumption of biomass and its impact on global land use (O'Brien et al. 2015, p. 242). A sustainable European bioeconomy, therefore, would need to engage the problem of an unsustainably high level of biomass consumption along with developing other, cooperative forms of transnational relations.

References

Babones, S.J. (2005). The Country-Level Income Structure of the World-Economy. *Journal of World-Systems Research, 11*(1), 29–55.

Backhouse, M., Lorenzen, K., Lühmann, M., Puder, J., Rodríguez, F., & Tittor, A. (2017). Bioökonomie-Strategien im Vergleich: Gemeinsamkeiten, Widersprüche und Leerstellen. *Working Paper 1, Bioeconomy & Inequalities,* Jena. https://www.bioinequalities.uni-jena.de/sozbemedia/neu/2017-09-28+workingpaper+1.pdf. Accessed 17 Dec 2018.

Bringezu, S., O'Brien, M., & Schütz, H. (2012). Beyond Biofuels: Assessing Global Land Use for Domestic Consumption of Biomass: A Conceptual and Empirical Contribution to Sustainable Management of Global Resources. *Land Use Policy, 29*(1), 224–232.

Bruckner, M., Häyhä, T., Giljum, S., Maus, V., Fischer, G., Tramberend, S., et al. (2019). Quantifying the Global Cropland Footprint of the European Union's Non-Food Bioeconomy. *Environmental Research Letters, 14*(4), 1–12.

Camia, A., Robert, N., Jonsson, R., Pilli, R., García-Condado, S., López-Lozano, R., et al. (2018). *Biomass Production, Supply, Uses and Flows in the European Union: First Results from an Integrated Assessment.* Luxembourg: Publications Office of the European Union. https://publications.jrc.ec.europa.eu/repository/bitstream/JRC109869/jrc109869_biomass_report_final2pdf2.pdf. Accessed 6 Nov 2020.

European Commission. (2012). *Innovating for Sustainable Growth: A Bioeconomy for Europe.* Luxembourg: Publications Office of the European Union. https://op.europa.eu/en/publication-detail/-/publication/1f0d8515-8dc0-4435-ba53-9570e47dbd51. Accessed 6 Nov 2020.

European Commission. (2018). *A Sustainable Bioeconomy for Europe: Strengthening the Connection Between Economy, Society and the Environment.* Brussels. https://ec.europa.eu/transparency/regdoc/rep/1/2018/EN/COM-2018-673-F1-EN-MAIN-PART-1.PDF. Accessed 6 May 2021.

Eurostat. (2018). Economy-wide Material Flow Accounts: Handbook 2018 edition (Manuals and guidelines). Luxembourg: Publications Office of the European Union. https://ec.europa.eu/eurostat/documents/3859598/9117556/KS-GQ-18-006-EN-N.pdf/b621b8ce-2792-47ff-9d10-067d2b8aac4b. Accessed 6 Nov 2020.

German Bioeconomy Council. (2015). Synopsis of National Strategies Around the World: Bioeconomy Policy (Part II). Berlin. https://biooekonomierat.

de/fileadmin/international/Bioeconomy-Policy_Part-II.pdf. Accessed 28 Oct 2019.

German Bioeconomy Council. (2018). Update Report of National Strategies Around the World: Bioeconomy Policy (Part III). Berlin. http://bioookono mierat.de/fileadmin/Publikationen/berichte/GBS_2018_Bioeconomy-Strate gies-around-the_World_Part-III.pdf. Accessed 22 June 2018.

Gudynas, E. (2011). Neo-Extraktivismus und Ausgleichsmechanismen der progressiven südamerikanischen Regierungen. *Kurswechsel, 3*, 69–80. http://www.beigewum.at/wordpress/wp-content/uploads/Neo-Extraktiv ismus.pdf. Accessed 6 Nov 2020.

Komlosy, A. (2016). Prospects of Decline and Hegemonic Shifts for the West. *Journal of World-Systems Research, 22*(2), 463–483.

Kovacs, B. (2015). Sustainable Agriculture, Forestry and Fisheries in the Bioeconomy: A Challenge for Europe. 4th SCAR Foresight Exercise. Luxembourg: Publications Office of the European Union. https://op.eur opa.eu/en/publication-detail/-/publication/7869030d-6d05-11e5-9317-01a a75ed71a1. Accessed 6 Nov 2020.

O'Brien, M., Schütz, H., & Bringezu, S. (2015). The Land Footprint of the EU Bioeconomy: Monitoring Tools, Gaps and Needs. *Land Use Policy, 47*, 235–246.

Piotrowski, S., Carus, M., & Essel, R. (2015a). Global Bioeconomy in the Conflict Between Biomass Supply and Demand (Nova Paper on Bio-Based Economy 7). Huerth. http://bio-based.eu/nova-papers. Accessed 30 May 2019.

Piotrowski, S., Essel, R., Carus, M., Dammer, L., & Engel, L. (2015b). Nachhaltig nutzbare Potenziale für Biokraftstoffe in Nutzungskonkurrenz zur Lebens- und Futtermittelproduktion, Bioenergie sowie zur stofflichen Nutzung in Deutschland, Europa und der Welt. Hürth. http://bio-based.eu/downloads/nachhaltig-nutzbare-potenziale-fuer-biokraftstoffe-in-nutzungskonkurrenz-zur-lebens-und-futtermittelproduktion-bioene rgie-sowie-zur-stofflichen-nutzung-in-deutschland-europa-und-der-welt/. Accessed 17 April 2019.

Priefer, C., Jörissen, J., & Frör, O. (2017). Pathways to Shape the Bioeconomy. *Resources, 6*(1), 1–23.

Ramcilovic-Suominen, S., & Pülzl, H. (2018). Sustainable Development—A 'Selling Point' of the Emerging EU Bioeconomy Policy Framework? *Journal of Cleaner Production, 172*, 4170–4180.

Ronzon, T., Lusser, M., Klinkenberg, M., Landa, L., Sanchez Lopez, J., M'Barek, R., et al. (2017). Bioeconomy Report 2016. JRC Scientific and Policy Report. Luxembourg: Publications Office of the European Union. https://ec.europa.eu/food/sites/food/files/safety/docs/fw_lib_swp_jrc-bioeconomy-report_2016.pdf. Accessed 6 Nov 2020.

Scarlat, N., Dallemand, J.-F., Monforti-Ferrario, F., & Nita, V. (2015). The Role of Biomass and Bioenergy in a Future Bioeconomy: Policies and Facts. *Environmental Development, 15*, 3–34.

Schaffartzik, A., & Pichler, M. (2017). Extractive Economies in Material and Political Terms: Broadening the Analytical Scope. *Sustainability, 9*(7), 1–17.

Svampa, M. (2012). Bergbau und Neo-Extraktivismus in Lateinamerika. In FDCL & RLS (Eds.), *Der Neue Extraktivismus: Eine Debatte über die Grenzen des Rohstoffmodells in Lateinamerika* (pp. 14–21). Berlin: FDCL-Verlag. https://www.fdcl.org/wp-content/uploads/2012/02/Der_Neue_Extraktivismus_web2.pdf. Accessed 6 Nov 2020.

UNEP. (2015). International Trade in Resources: A Biophysical Assessment: Report of the International Resource Panel. Nairobi: United Nations Environment Programme. https://www.resourcepanel.org/sites/default/files/documents/document/media/-international_trade_in_resources_full_report_english_0.pdf. Accessed 6 Nov 2020.

Wallerstein, I.M. (2007). *World-Systems Analysis: An Introduction* (5th ed.). Durham: Duke University Press.

14 Sustaining the European Bioeconomy ...

Open Access This chapter is licensed under the terms of the Creative Commons Attribution 4.0 International License (http://creativecommons.org/licenses/by/4.0/), which permits use, sharing, adaptation, distribution and reproduction in any medium or format, as long as you give appropriate credit to the original author(s) and the source, provide a link to the Creative Commons license and indicate if changes were made.

The images or other third party material in this chapter are included in the chapter's Creative Commons license, unless indicated otherwise in a credit line to the material. If material is not included in the chapter's Creative Commons license and your intended use is not permitted by statutory regulation or exceeds the permitted use, you will need to obtain permission directly from the copyright holder.

15

Towards an Extractivist Bioeconomy? The Risk of Deepening Agrarian Extractivism When Promoting Bioeconomy in Argentina

Anne Tittor

15.1 Introduction: Argentina as a Bioeconomy Pioneer

Argentina considers itself a Latin American pioneer when it comes to the bioeconomy. The debate on bioeconomy in Argentina is broader and more intense than in other countries, especially among others in the Global South. Since 2013, two national and ten regional conferences[1] on bioeconomy have been held in the country, at which government representatives, researchers, actors from business and agrarian organizations

---

[1]The national conferences took place between 21 and 22 March 2013 and 5 and 6 June 2014 in Buenos Aires. The Regional Conferences in 2015 were held between 16 and 17 April in Puerto Madryn, 6 and 7 May in Posadas, 25 and 26 June in Rosario and 1 and 2 July in Tucuman. In 2016, regional symposia were held between 22 and 23 September in Cuyo, 4 and 5 October in Córdoba, 20 and 21 October in Resistencia, 17 and 19 November in Neuquén, 24 and 25 November in Salta and 11 and 13 December in Buenos Aires.

A. Tittor (✉)
Friedrich-Schiller-Universität Jena, Jena, Germany
e-mail: anne.tittor@uni-jena.de

© The Author(s) 2021
M. Backhouse et al. (eds.), *Bioeconomy and Global Inequalities*,
https://doi.org/10.1007/978-3-030-68944-5_15

discussed how to establish a bioeconomy that would reflect Argentina's needs and interests. A total of between 4000 and 5000 people attended these conferences (key bioeconomy policy maker, Interview no. 1).

Bioeconomy simultaneously promises sustainability, innovation and competitiveness, and promotes bioprocesses, bioproducts and biofuels. This implies increasing demand for biomass and, therefore, a huge impact on the sector producing it: primarily the agricultural sector.[2] Nevertheless, Argentina's agricultural policies have been contested for several years, and the expansion of soybean monocultures in the last three decades and the related use of pesticides in particular (Teubal and Giarracca 2013a).

Against this background, this chapter asks which kind of agriculture is promoted within the bioeconomy debate in Argentina. More precisely, it examines the motivation and objectives presented by Argentinian advocates of the bioeconomy and the issues that are marginalised and excluded from the debate. It also looks at the role played by the actors that currently promote bioeconomy in Argentina within the transformation of agriculture that has taken place over the last few decades.

If bioeconomy involves agriculture delivering more than just food and fodder (it's role in the past), and instead includes the provision of additional raw materials for bioenergy such as biogas, biomaterials and bioplastics (where in most, if not all, cases, oil-based resources were used), the demand for biomass rises considerably (see also Lühmann in this volume). Therefore, a critical approach to bioeconomy should ask who aims to produce this additional biomass, under which conditions and at which social and environmental cost. Is there a danger that countries in the Global South will (once again) be pushed purely into the role of producers of resources that are intended for export, while destroying and exhausting local livelihoods and nature? This core question has been at the focus of recent debates about extractivism.

---

[2] Besides the agricultural sector, there are efforts to produce biomass using algae. These innovations, as well as other aspects of what is known as the "blue bioeconomy", are not addressed in this chapter. The same applies to the "white bioeconomy"—the medical field.

Therefore, for my discussion of Argentina's bioeconomy, I refer to considerations from the Latin American centred debate about extractivism (Svampa 2012; Gudynas 2015), especially agrarian extractivism (McKay 2017), to describe the changes in Argentina's agriculture. This serves as a background to evaluate the kind of changes that the bioeconomy will bring about, and which structures will remain the same.

My analysis draws on material from several conferences and workshops on bioeconomy (listed in footnote 1), policy documents, media reports, material from web-based further education programmes and the websites of key actors. It takes into consideration the statements of the Argentinian delegation at the Global Bioeconomy Summit in Berlin 2018 as well as expert interviews with key actors conducted during field research in May and June 2018 in Buenos Aires and Córdoba (listed at the end). It further draws on the broad scientific debate in Argentina about the transformation of agriculture over the last few decades and the related problems, a debate, which I will briefly outline in the following section. Using this framework, Sect. 15.3 provides some information about the impacts of soybean expansion over the last few decades. Against this background, in Sect. 15.4 I discuss the extent to which the Argentinian debate about bioeconomy relates to the problematic impacts of soybean monocropping and the proposals to overcome them. In Sect. 15.5, I conclude that, despite some counter-tendencies, implementing the current understanding of bioeconomy in Argentina risks deepening agrarian extractivism.

15.2 Agrarian Extractivism as a Tool for Analysing Argentina's Bioeconomy

Over the last few years, the notion of extractivism has been widely discussed to characterize an economic model or accumulation strategy that relies on the extraction of raw materials, especially from mining activities in the Global South, and their export towards the Global North (Svampa and Viale 2014; Bebbington et al. 2014; Gudynas 2015). Although extractivism dates back to the colonial era, in the twenty-first century a re-primarization of the economies in Latin America took place, which, in times of high prices for raw materials on the world market, became an important driver of growth and economical dynamics. In 2012, Maristella Svampa argued that extractivism is not limited to mining, but can also include the agricultural sector when it is organized as monocultural, intensive farming, and is based on the over-exploitation of soils and ecosystems. This form of farming leads to an increase in the use of pesticides, herbicides and fertilizer and tends to benefit transnational and national elites. It causes local environmental damage, severe health impacts for the local population due to spraying of increasing amounts of pesticides, and the displacement and dispossession of local communities (Svampa 2012, p. 15). This type of agriculture is purely oriented towards global markets and competitiveness, not towards food sovereignty. Eduardo Gudynas (2015) uses the term "agricultural extractivism" to describe the trend towards monoculture with little or no processing, the use of transgenic crops, heavy machinery, chemical herbicides and the export of the product as a commodity.

Miguel Teubal and Norma Giarracca (2013a, p. 15) also view extractivism as connected to re-primarization and argue that there are many parallels between opencast mining and agribusiness. They define the re-primarization of the economy as the strengthening of primary products for export, which are treated as commodities, including those produced by the agribusiness and agro-industry sector for the internal market. These sectors can be highly capital intensive, generate huge profits, but employ few people and retain the characteristics of an economic enclave. They are based on the over-exploitation and exhaustion of natural resources such as water and soil, and imply environmental degradation and the depletion of human and animal health. Whereas for tens of thousands of years, agriculture did not exhaust its conditions of existence, today's practices are different. State-of-the-art technologies are promoted, such as biotechnology, information technology and nanotechnology, but normally depend on heavy and increasing pesticide use. These activities consume high quantities of non-reproducible resources such as fertile soil and biodiversity (Teubal and Giarracca 2013b, p. 23). They displace rural workers and peasants, and cause conflict with other land uses. They are not essential for local communities and do not fulfil basic needs as the commodities they produce are for export and global markets.

Soil (exhaustion) is a major issue. The UN declared 2015 the year of soils to draw attention to the necessity of protecting soils, which the UN Food and Agricultural Organisation FAO classifies as a non-renewable resource.[3] What at first seems contradictory, as soil has recovered by itself over centuries, is the outcome of an intensified form of agriculture that is continuously extracting nutrients from the soil. To counterbalance extraction, synthetic fertilizers are used more and more—about half of all synthetic fertilizers used in the history of mankind have been used in the last 25 years (UNEP 2014, p. 14). Virginia Toledo López argues that there are two different notions of agrarian extractivism (2017, p. 2), one articulated by Pengue and another by Gras and Hernandez: "mining agriculture" (Pengue 2005a, p. 19) depends on fossil inputs as it extracts too many nutrients from the soil, with soybean being one of the worst crops in this respect. This form of agriculture reinforces classical extractivism, and even with the required inputs, the quality of soils continues to worsen. In contrast, Gras and Hernandez (2013) insist that the agribusiness model itself, regardless of the type of crop or circumstances, is a form of production, commodification and organization that intensifies the extractivist features already present in agriculture.

Based on his analysis of Bolivia's agriculture, Ben McKay (2017) proposes defining agrarian extractivism as a combination of four interlinked dimensions: firstly, large volumes of materials are extracted and destined for export with little or no processing, which fuels industrialization in the Global North. Second, the value chain is highly concentrated and controlled by a small number of enterprises, combined with sectoral disarticulation, which refers to its lack of linkages to other sectors of the economy (McKay 2017, p. 204). As a third element, McKay identifies a high intensity of environmental degradation including over-exploitation of soils, contamination of water sources, a loss of biodiversity and deforestation. Lastly, agrarian extractivism involves the deterioration of labour opportunities and/or labour conditions, as people are dispossessed of their lands and lack job opportunities. In the context of soybean, McKay

---

[3] "The large difference between erosion rates under conventional agriculture and soil formation rates implies that we are essentially mining the soil and that we should consider the resource as non-renewable. [...] However, long-term sustainability requires that soil erosion rates on agricultural land are reduced to near-zero levels" (FAO 2015, p. 103).

stresses that monocultural cropping in an industrial form only offers very few jobs (ibid., p. 208).

However, McKay only sees displacement in terms of job opportunities, not in terms of land rights or violent conflict. Nevertheless, these factors also deserve to be treated as a dimension on their own. At the same time, McKay also concentrates on exports. However, other authors have argued (at least in the Argentinian context) that the production of agrofuels can also be interpreted as a form of extractivism, because it involves an accumulation mechanism based on the appropriation and exploitation of natural resources that intensifies the expansion of soybean (Toledo López 2013, p. 157). Furthermore, agrofuels—even if they are produced for local and national use—strengthen the agribusiness model and the power of large enterprises while weakening peasant agriculture and food sovereignty (Teubal and Giarracca 2013b, p. 28).

In summary, I propose the following understanding of agrarian extractivism: agrarian extractivism is an accumulation strategy based on a model of land use that is established by agribusiness actors to profit from the monoculture of certain crops that are treated as commodities. Agrarian extractivism is based on the expulsion of peasants and indigenous groups by force and/or by market mechanisms from their traditional lands. The crops are planted to achieve the highest possible profits by selling them on a global food, fodder, fuel or energy market. Often, the crops are produced for export, but even if they are not they do not serve as food for local communities. Agrarian extractivism is based on the over-exploitation of soils and the permanent extraction of nutrients, which agribusiness seeks to compensate for by increasing the use of fertilizers, which are normally based on fossil fuels. Agrarian extractivism often goes hand in hand with strengthening large business actors, reinforced sectorial disarticulation, re-primarization and the emergence of structures typical of enclave economies. It is oriented towards the use of modern and digital technologies and machinery and contributes to an agriculture without peasants. Finally, its externalities are the destruction of soils, the contamination of water sources, a loss of biodiversity, deforestation and land concentration.

15.3 The Expansion of Soybean as Agrarian Extractivism in Argentina

In the last few decades, the expansion of soybean has led the agricultural sector in Argentina to undergo tremendous change. This dynamic holds all of the characteristics of agricultural extractivism: today, soybean is the most important single crop in the country; it has transformed landscapes, social structures and even the entire economy (Lapegna 2016). In 2010, 18.5 million hectares (ha) were used for soybean production, which represented 58% of all cultivated lands in the country (Ainsuain and Echaguibel 2012, p. 93). Only one million ha (which produced 3 million metric tons of soybean) were dedicated to feeding animals in Argentina—94% of the crop was directly exported. In 1996, Argentina became one of the first countries to permit the cultivation of GMO soybean and corn (Mikkelsen 2008, p. 171). This is referred to as "the technological package" in Argentina and it also led to the dominance of one of the biggest agricultural companies in the world, Monsanto, which holds patents on the genetically modified soybeans seeds that are resistant to the glyphosate-based pesticide Roundup. Additionally, soybean monoculture is based on a specific form of farming (no till) and a form of organizing work, which is non-intensive, but highly specialized. Furthermore, it contributes to the restructuring of the agricultural sector concerning the importance of agribusiness actors, the diminution of crop rotation, pasture farming and crop diversity (Gras and Hernandez 2013, p. 29). Moreover, transnational corporations have gained control over the sector: in 2002 only 6 companies processed 92% of soy flour, and seven TNCs controlled about 60% of exports (Lapegna 2016, p. 32). The value chain is highly concentrated, controlled by very few enterprises, and characterized by sectoral disarticulation, as soybean production lacks linkages with other sectors of the economy.

The socio-environmental impacts are huge: whereas in the 1990s, soybean monoculture was displacing cattle ranging in the Pampa, it has since expanded into more fragile ecosystems in the North-East of the country, including the temperate rain forests of the Yungas and El Chaco where it contributes to deforestation (Pengue 2005b, p. 315; Fehlenberg et al. 2017). With the expansion of GMOs and especially soybean, the quantity of pesticides used skyrocketed. In 1996, 39 million litres equivalent in kilos of glyphosate were used in Argentina; by 2015, this level had increased to 369 million litres. Argentina has the highest level of exposure to glyphosate per person and per year in the world (Avila-Vazquez and Difilippo 2016, p. 23). This impacts heavily on human health. However, it is not only pesticide use, but also the use of fertilizer that has increased. Before 1990, most soils in the Pampa had yet to be fertilized. This changed dramatically with the cultivation of soybean,[4] which played an emblematic role in the loss of Argentinian soil quality (Pengue 2005b, p. 315). Soybean is one of the crops that extracts the most from soil per unit production (Pengue 2015, p. 13). The focus in Argentina on soybean contributes significantly to the loss of soil quality. The country, therefore, exports its nutrients, and even its soil. This causes ecological debt, which risks the basis of future agriculture in Argentina (ibid., p. 14). Argentina has one of the highest rates of extraction per person in terms of material flows in Latin America (16,46 t/per capita).

The implementation of the "technological package" contributed to the possibility of cultivating soybean on soils where this had seemed impossible for a long time, and to the fact that the crop expanded towards North-East Argentina, where it displaces indigenous communities and causes conflicts over land (Toledo López 2016, p. 197; REDAF—Red Agroforestal Chaco Argentina 2013). Even those who promote soybean admit that it currently only generates 197,000 jobs, which is around 10% of the jobs in all agri-food value chains, while occupying a full 58% of agricultural land (Bragachini 2011). Although optimists expect that each direct job in this industry will generate 3.8 additional jobs in other sectors (Llach et al. 2004), they fail to account for the lost jobs due to the change in land use.

---

[4]In terms of material flows, this indicates a large change: in the 1990s, the country began exporting a considerable amount of nutrients each year—among them nitrogen and phosphorus. The country exports around 3,500,000 metric tons of nutrients a year as intensification does not allow for a process of natural replenishment to occur and means a huge loss in the long term (Pengue 2005b, p. 320).

15.4 Argentina's Expectations for the Bioeconomy

These theoretical considerations on agrarian extractivism and this short outline of current tendencies within Argentinian agriculture form the background against which to assess the shape of the Argentinian bioeconomy. I place special emphasis on whether and how bioeconomy advocates position themselves towards GMO crops and soybean expansion. Furthermore, I examine the expectations placed on the bioeconomy to achieve a more socio-environmentally friendly form of agriculture.

Most documents on bioeconomy in Argentina start by stating that the concept was developed in Europe and in industrialized countries in particular (Guy et al. 2014). Bioeconomy has been discussed in Argentina since about 2013 and first used the framing present in many papers by the OECD and the European Union. As such, the bioeconomy was viewed as an answer to the challenges of population growth, climate change and the importance of overcoming the reliance on fossil fuels (MINCyT—Ministerio de Ciencia, Technología e Innovación Productiva n.d.). Over the years, Argentina has begun appropriating and developing its own interpretation of the concept, which can be defined as 1) a strong reliance on biotechnology and no-till farming, including the use of large amounts of fertilizers and pesticides; 2) the intention to gain "added value" and 3) a reliance on a certain number of sustainable innovations.

15.4.1 Biotechnology, Fertilizers, Pesticides and no-till Farming as a Key Basis of Bioeconomy

Bioeconomy is seen as a great opportunity for Argentina due to good pre-conditions in terms of resources and agricultural, industrial and economic structures (MINAGRO—Ministerio de Agroindustria 2016). The advocates of bioeconomy in Argentina are not focused on a particular crop, such as soybean, but on the opportunity to produce biomass as the key resource for economic progress. Part of the argument includes the claims that Argentina is a pioneer in GMO crops, and that there is broad acceptance for biotechnology within society (ibid.). Argentina has (per capita) the highest number of biotechnology companies and researchers in Latin America (ProsperA. Investment Opportunities 2009). This narrative is based on the assumption that the environmental impact of agriculture is per se diminished by biotechnology, as the following quote from the Inter-American Institute for Cooperation on Agriculture (IICA) shows:

> Argentina plays a very important leading role in biotechnology for genetically modified crops. By developing our own technology, we have increased our sector's productivity and reduced the environmental impact of our production activities. (IICA 2018)

The Secretary of Food and Bioeconomy chose these words to explain that Argentina should serve as a role model to develop the bioeconomy in Latin America precisely because of the country's early and complete introduction of GMOs and its efforts to support its own biotechnology laboratories. Biotechnology laboratories have significantly expanded over the last few years and have gained new buildings and more personnel than ever before. The promoters of bioeconomy, therefore, rely on a key pillar of agrarian extractivism: GMO crops, and soybean in particular. The same actors that promote GMOs and biotechnology are now promoting the bioeconomy. Therefore, it is no surprise that there is an underlying consensus within the bioeconomy debate in Argentina that agriculture heavily relies on biotechnology and pesticides. Both are seen as necessary to produce more crops for export and internal use, and to deepen intensive agriculture. There is no questioning of monoculture production and there is no criticism within the entire bioeconomy debate about the increasing pesticide and herbicide use. Despite several campaigns by social movements (Arancibia 2013; Carrizo and Berger 2014) and alarming studies (Avila-Vazquez et al. 2018; Verzeñassi 2014), the epistemic bioeconomy community is not interested in taking up these issues.

In addition to biotechnology, new farming techniques are often presented as the solution to environmental challenges. Those arguing for the bioeconomy stress the need for more environmentally friendly agricultural techniques—and often present no tillage farming as the key. No-till farming was introduced with soybean expansion in Argentina. Organizations that promote no-till farming, such as AAPRESID (Argentinian Association of no-till Farming), claim to introduce a form of sustainable agriculture that protects the soil and is based on knowledge and technological innovation. Nevertheless, the scientific debate shows that the extraction of nutrients from soils and soil degradation continue with no-till farming. The soil report states:

> Despite the wide adoption of no-tillage, intensive annual cultivation (largely of soybean) and the lack of rotation with other crops or pastures have resulted in soil degradation by wind and water erosion, waterlogging, compaction, sealing/capping, and soil fertility depletion. (FAO 2015, p. 368)

The issue as to whether no-till farming significantly increases soil organic stocks and how much it helps to improve the physical properties of topsoil is controversial (FAO 2015, p. 383). However, the evidence suggests that no-till farming has yet to reverse the process of soil degradation (Pengue 2015, p. 13). Moreover, as long as monocropping continues, a change of direction is unlikely. The promoters of Argentinian bioeconomy do not have any problems with monoculture, as it represents a further pillar of agrarian extractivism.

15.4.2 Agro-Industrialization and "Adding Value" as a Key Goal Within Bioeconomy

In the understanding of its proponents in Argentina, bioeconomy includes an opportunity to "add value" to agricultural products. This idea is also expressed by the Argentinian Stock Exchange:

> The impulse of the bioeconomy is extremely attractive in countries of Latin America, in which the increase of added value to the agricultural

primary production appears to be of crucial importance for the development of their respective economies. (Bolsa de Cereales and Wierny 2015)

Therefore, factories and installations that transform agricultural products into biofuels or biomaterials are seen as key contributors to the bioeconomy. This narrative is reproduced by a manager of an ethanol factory in the province of Córdoba (bioethanol factory manager, Interview no. 2). The strategy of "adding value" is nothing new and was dominant during the expansion of soybean monoculture (Bernhold 2019). Norma Giarracca and Miguel Teubal identify the promise of "agro-industrialization" and the "incorporation of value" as key elements put forward by proponents of agrarian extractivism (Teubal and Giarracca 2013a, p. 10). For example, in 2014 the then-Argentinian president, Cristina Fernández de Kirchner (who held office between 2007 and 2015) stressed:

> We need to add value to our products. The idea is to industrialize rurality, which means that primary products have added value. Therefore, it is necessary to invest in research and development [...] fighting inequality means generating jobs and not allowing our exports to continue to be re-primarised. (Kirchner 2014)

The idea of "added value" became hegemonic under Kirchner's successor Mauricio Macri (in office between 2015 and 2019). However, neither Macri and his constituency nor the Kirchnerists are seriously concerned about the environmental impact of intensive monocultural crops.

Added value is often proposed as a counter-tendency to re-primarization and, therefore, as contributing to overcoming another pillar of agrarian extractivism. Several of my interview partners stressed that Argentina does not want to play the role of the breadbasket, or export its agricultural products, but wants to industrialize its agriculture as this is seen as a way to generate jobs in the countryside, far from the urban centres (key bioeconomy policy maker, Interview no. 1). Some argue, that doing so will even open up new jobs for high-skilled labour in the countryside and provide young people with new opportunities (bioethanol factory manager, Interview no. 2). Many of my interview partners underline that Argentina wants a bioeconomy with social inclusion. The only form of inclusion they imagine, however, involves industrializing agricultural jobs and providing start-ups with the conditions they need to expand. Nevertheless, only a limited number of people would benefit from such policies and it is clear that most peasants, indigenous groups and landless rural workers will not do so. My interview partners argued that new entrepreneurs implement the bioeconomy on the ground. Some individuals from business support small social projects f.e. projects of urban agriculture—but this merely involves small-scale corporate social responsibility and does not involve making changes to the way in which their businesses work. Key policy makers—including those who promote bioeconomy as a development strategy—never mention other forms of participation or inclusion.

When presented as a development strategy, proponents often stress decentral units of production and the necessity to "add value" very close to the site of production:

> The fact that biomass does not travel well represents an advantage for the issues connected to bioeconomy. It means that we have to opt for, or at least focus on, local development. We need to transform our surroundings as much as possible, and build the highest number of industrial plants that we can for energetic use and biomolecular enterprises etc. This leads to a focus on local systems and probably on decentralization. (key bioeconomy advisor, Interview no. 3)

The quoted advisor and others continuously stress that "biomass does not travel well", which means that it is not profitable to transport biomass over long distances. This could potentially lead to a change in agro-industrial patterns: soybean is currently transported to the next port for export, or to mills and refineries on the way to the port. In order to counteract this tendency and to improve profitability, a lot of new infrastructure needs to be built, and processing will have to be implemented on a large scale. This will have to happen if the argument that the bioeconomy represents a counter-tendency to export orientation is ever to be taken seriously. It is even more difficult to build a counterweight to re-primarization; industrialization grounded on bio-based products, therefore, will probably both strengthen the agricultural and the industrial sector at the same time so that the relative weight of exports from mining and agriculture diminishes. Nevertheless, as long as no or very little processing takes place, which, according to CEPAL[5] categorization, includes vegetable oils and agrofuels (Gudynas 2015, p. 16), and the products continue to be sold as commodities, industrialization still contributes to re-primarization and the deepening of agrarian extractivism—even if the former president believes the opposite to be the case (Teubal and Giarracca 2013a, p. 11).

The issue of infrastructure is not seriously discussed, nor is the issue of land distribution addressed, although soybean expansion has contributed to an immense concentration of land ownership and led many farmers to give up agriculture. Additionally, in northern areas of the country, conflicts over land continue because of the expanding soybean and commodity frontiers, which are ignored by the debate entirely (REDAF 2013).

15.4.3 On Sustainable Innovations and Counter-Tendencies to Agrarian Extractivism

Despite the continuity in terms of actors and policy orientations between those promoting soybean and those pushing for the bioeconomy and, by implication, for a deepened agrarian extractivism, there are also some new discussions and interesting proposals that potentially contribute to more sustainable patterns of production and consumption.

First, one additional focus of the bioeconomy debate is the use of waste products as biomass. At the Córdoba symposium, a debate took place about the possibility of generating energy from meat waste. Pilot projects within the bioeconomy framework are testing how much energy can be obtained from slaughterhouse waste to produce electricity

---

[5]CEPAL is the Spanish acronym for Economic Commission for Latin America and the Caribbean.

and heat as well as the use of bovine blood for several products with "added value" (Manfredi and Kalbermatter 2016). In the regions of Cuyo and Mendoza, many projects are working on using by-products from different agricultural production chains.[6] Determining which waste products are made by different agro-industrial businesses has helped promote the re-use of by-products and to encourage cascade use. In the most recent bioeconomy symposia (1 and 2 November 2019 in Patagonia) issues were raised for the first time about food sovereignty, the contribution of indigenous knowledge about food, and female subsistence work.

Second, environmental issues are gaining importance within the Argentinian business community. For example, managers and owners of bioenergy enterprises have begun representing their enterprises as sustainable (Toledo López and Tittor 2019). They claim to be concerned about environmental issues and to seek answers within the framework provided by ecological modernization. The proponents of bioenergy and biofuel production have begun studying the CO_2 emissions linked to their production processes. On the one hand, this is essential if they are to gain access to the international market. However, sometimes they discuss emission reductions that go beyond those required by international treaties (Hilbert 2016), and seriously consider issues such as the cascade use of their materials, avoiding waste and saving energy. When Bio4 realised that their production of bioethanol produced more corn by-products than farmers in the region required as feed, they constructed a bioelectricity plant next to the factory that now produces the energy needed by the enterprise (two managers from both companies, Interview no. 4 and 5). Even strong supporters of the soybean model admit that more efforts are necessary to improve environmental performance. As such, there is at least a level of recognition that a problem exists, although the solution proposed is certification of no-till farming and crop rotation. Moreover, advocates admit that although large-scale farmers' organizations always recommend crop rotation, only about 30% of farmers

---

[6] See 11 presentations on the bioeconomy symposium in Cuyo between 22 and 23 September 2016.

actually follow these recommendations (key bioeconomy advisor, Interview no. 3). This is because monocropping is more profitable, at least in the short-term: on average, soybean monoculture provides about 90% higher profits than a rotation system (Peretti 2013, p. 33).

Third, there are interesting and innovative small-scale projects developing as part of the bioeconomy framework; the aforementioned example of Bio4/Bioelectrica is one of them. Integrating rice cultivation and fish production, where waste from one product is used to produce the other, represents a further "success story". In this case, a pilot programme that contributes to the development of a form of rice production that forgoes the need for fertilizers and pesticides is often presented as a role model for further bioeconomy projects (bioeconomy expert, Interview no. 6).

15.5 Conclusion: Towards an Extractive Bioeconomy?

This chapter has shown that Argentina's bioeconomy has a clear agro-industrial and bio-technological focus. In Argentina, bioeconomy is framed as meaning further intensification of agro-industrial production—including GMOs and the immense use of pesticides—combined with strengthening industrial upgrading. The same people and institutions that have supported soybean expansion over the last few decades also advocate bioeconomy.

Environmental issues are mentioned, but GMO-seed, bio-technological processes and no-till farming are presented as key strategies with which to increase sustainability and reduce environmental impact. There are no measures to reduce pesticide use or monocultural commodity cropping. Argentina's bioeconomy is clearly a growth strategy; nature conservation and sustainability do not play central roles in this policy. Although agroecological initiatives have gained momentum within Argentina, they do not form part of the bioeconomy debate, and there is very little communication between these different epistemic communities (on one exception to this, see

Arancibia 2013). However, the recent symposium in Patagonia indicates that this could change in the future.

Bioeconomy in Argentina does not involve a rupture with GMO crops, intensive herbicide and pesticide use or monocultures, and it continues to be strongly oriented towards global markets. Agroecological initiatives, a solidarity economy or de-centralized energy systems are not part of the debate. This form of bioeconomy has huge advantages for agricultural producers of commodity crops, biofuel producers and biotech labs. However, there is no strategy for including people such as peasants or indigenous groups in this model of development. The only form of inclusion is a vague promise of jobs due to the construction of agro-industrial installations as part of upgrading. The demands made by ecological movements in Argentina (such as to stop spraying, or to end deforestation and for environmental justice; see Merlinsky 2020) are not even recognized or addressed. Nor are the struggles against the dispossession of indigenous communities and small peasants acknowledged—processes that are often driven by agro-industrial expansion into new territories.

Argentina has not merely adopted the framing of bioeconomy by the OECD countries, but has appropriated the debate and developed its own understanding of the concept. I propose the term "extractivist bioeconomy" as characterizing this form of bioeconomy. At first glance, "extractivist bioeconomy" appears to be an oxymoron, as a bioeconomy, by definition, relies on biological processes and is aimed at sustainability. Nevertheless, there is no automatism to sustainability or to remaining inside planetary boundaries. Instead, bioeconomy in Argentina actually deepens the extractivist tendency within the dominant form of agriculture by exhausting soils due to monocropping and, therefore, deepens agrarian extractivism. As a means of overcoming soil exhaustion, pests and other problems, more biotechnology and more pesticides are said to be required. Nevertheless, these pesticides contaminate the water, ecosystems and the people who work, live and go to school near such plantations. Finally, even if the tendency towards re-primarization can be overturned through industrial upgrading and investment in infrastructure, the socio-environmental problems associated with a highly intensified form of agriculture based on monocropping, biotechnology and pesticides will remain.

List of Interviews quoted

Interview no.	Profession/function	Date and place
Interview no. 1	Key bioeconomy policy maker	7 June 2018, Buenos Aires
Interview no. 2	Bioethanol factory manager	12 June 2018, Córdoba
Interview no. 3	Key bioeconomy advisor	4 June 2018, Buenos Aires
Interview no. 4	Bioethanol factory manager	15 June 2018, Rio Cuarto
Interview no. 5	Bioelectricity factory manager	15 June 2018, Rio Cuarto
Interview no. 6	Bioeconomy expert	8 December 2017, Berlin

References

Ainsuain, O., & Echaguibel, M. (2012). *A 100 años del Grito de Alcorta. Soja, Agronegocios y explotación*. Buenos Aires: Ediciones Ciccus.

Arancibia, F. (2013). Challenging the Bioeconomy: The Dynamics of Collective Action in Argentina. *Technology in Society, 35*, 70–92.

Avila-Vazquez, M., & Difilippo, F.S. (2016). Agricultura tóxica y salud en pueblos fumigados de Argentina. *Crítica y Resistencias. Revista de conflictos sociales latinoamericanos* (2), 23–45.

Avila-Vazquez, M., Difilippo, F.S., Lean, B.M., Maturano, E., & Etchegoyen, A. (2018). Environmental Exposure to Glyphosate and Reproductive Health Impacts in Agricultural Population of Argentina. *Journal of Environmental Protection, 9*, 241–253.

Bebbington, A., Cuba, N., & Rogan, J. (2014). The Overlapping Geographies of Resource Extraction. *Revista. Harvard Review of Latin America, 13*(2), 20–24.

Bernhold, C. (2019). *Upgrading and Uneven Development: On Corporate Strategies and Class Dynamics in Argentinian Grain and Oilseed Value Chains*. Dissertation. Zurich: University Zurich.

Bolsa de Cereales, & Wierny, M. (2015). *Medición de la bioeconomía. Cuantificación del Caso Argentino.* Buenos Aires: Libro digital.

Bragachini, M. (2011). La cadena de la soja, en 10 años, puede generar 400 mil puestos de trabajo. Infocamp. https://www.infocampo.com.ar/bragac hini-la-cadena-de-la-soja-puede-generar-400-000-puestos-de-trabajo-en-10-anos/. Accessed 30 Feb 2020.

Carrizo, C., & Berger, M.S. (2014). Luchas contra los pilares de los agronegocios en Argentina: transgénicos, agrotóxicos y CONABIA. *Letras Verdes, Revista Latinoamericana de Estudios Socioambientales,* 4–28.

FAO. (2015). *Status of the World's Soil Resources: Main Report.* Rome: FAO; ITPS.

Fehlenberg, V., Baumann, M., Gasparri, N.I., Piquer-Rodriguez, M., Gavier-Pizarro, G., & Kuemmerle, T. (2017). The Role of Soybean Production as an Underlying Driver of Deforestation in the South American Chaco. *Global Environmental Change, 45,* 24–34.

Gras, C., & Hernandez, V.A. (2013). Los pilares del modelo agribusiness y sus estilos empresariales. In C. Gras & V.A. Hernandez (Eds.), *El agro como negocio: Producción, sociedad y territorios en la globalización* (pp. 17–46, Sociedad). Buenos Aires: Biblos.

Gudynas, E. (2015). *Extractivismos. Ecología, economía y política de un modo de entender el desarrollo y la Naturaleza.* La Paz: Centro de Documentación e Información Bolivia (CEDIB).

Guy, H., Pahun, J., & Trigo, E. (2014). La Bioeconomía en América Latina: oportunidades de desarrollo e implicaciones de política e investigación. *Faces, 20*(42), 125–141.

Hilbert, J.A. (2016, Oct 4). Impactos medioambientales de la transformación de biomasa en energía y otros productos comercializables. Estudios de casos argentinos. Bioeconomy Simposium, Córdoba.

IICA. (2018). Argentina Will Become the American Continent's Hub for Knowledge in the Field of Bioeconomy. https://iica.int/en/press/news/arg entina-will-become-american-continent%25E2%2580%2599s-hub-knowle dge-field-bioeconomy. Accessed 30 March 2020.

Kirchner, C. (2014). Necesitamos agregar valor a nuestros productos. Celac. https://www.minutouno.com/notas/311897-cristina-kirchner-la-celac-necesi tamos-agregar-valor-nuestros-productos. Accessed 30 March 2020.

Lapegna, P. (2016). *Soybeans and Power: Genetically Modified Crops, Environmental Politics, and Social Movements in Argentina.* Oxford: Oxford University Press.

Llach, J., Harriague, M., & O'Connor, E. (2004). *La generación de empleo en las cadenas agroindustriales*. Buenos Aires: Fundación Producir Conservando.

Manfredi, M.J., & Kalbermatter, L. (2016, Oct 4). *Valorización de subproductos de la cadena cárnica regional*. Bioeconomy simposium, Córdoba, Argentina.

McKay, B. (2017). Agrarian Extractivism in Bolivia. *World Development, 97*, 199–211.

Merlinsky, G. (2020). *Cartografías del conflicto ambiental en Argentina III*. Buenos Aires: Ciccus.

Mikkelsen, C.A. (2008). La expansión de la soja y su relación con la agricultura industrial. *Revista Universitaria de Geografía, 17*(1), 165–188.

MINAGRO. (2016). Bioeconomía Argentina. Visión desde Agroindustria. www.agroindustria.gob.ar/sitio/areas/bioeconomia/_archivos//000000_Bioeconomia%20Argentina.pdf. Accessed 26 May 2020.

MINCyT. (n.d.). *El paradigma de la bioeconomía*. http://www.bioeconomia.mincyt.gob.ar/bioeconomia-argentina/. Accessed 19 May 2020.

Pengue, W.A. (2005a). *Agricultura industrial y transnacionalizacion en America Latina: ¿la transgenesis de un continente?* Buenos Aires: GEPAMA—Grupo de Ecología del Paisaje y Medio Ambiente. http://aao.org.br/aao/pdfs/publicacoes/agricultura-industrial-y-transnacionalizacion-en-america-latina.pdf. Accessed 26 May 2020.

Pengue, W.A. (2005b). Transgenic Crops in Argentina: The Ecological and Social Debt. *Bulletin of Science, Technology & Society, 25*, 314–322.

Pengue, W.A. (2015). Suelos, Huellas de Nutrientes y Estabilidad Ecosistémica. *Fronteras, 13*, 1–18.

Peretti, P. (2013). *Chacareros, soja y gobernabilidad. Del Grito de Alcorta a la Resolución 125*. Buenos Aires: Ciccus.

ProsperA. Investment Opportunities. (2009). Biotechnology in Argentina. Knowledge + Innovation to Meet Global Market Needs. https://www.assolombarda.it/fs/201029114832_122.pdf. Accessed 26 May 2020.

REDAF. (2013). *Conflictos sobre tenencia de tierra y ambientales en la región del Chaco argentino*. Reconquista/ Chaco: 3º Informe.

Svampa, M. (2012). Bergbau und Neo-Extraktivismus in Lateinamerika. In FDCL & RLS (Eds.), *Der Neue Extraktivismus. Eine Debatte über die Grenzen des Rohstoffmodells in Lateinamerika* (pp. 14–33). Berlin: FDCL-Verlag.

Svampa, M., & Viale, E. (2014). *Maldesarrollo. La Argentina del Extractivismo y el despojo* (2nd ed.). Buenos Aires: Katz Editores.

Teubal, M., & Giarracca, N. (2013a). Introducción. In N. Giarracca & M. Teubal (Eds.), *Actividades extractivas en expansión. Reprimarización de la economía argentina?* (pp. 9–18). Buenos Aires: Antropofagia.

Teubal, M., & Giarracca, N. (2013b). Las actividades extractivas en la Argentina. In N. Giarracca & M. Teubal (Eds.), *Actividades extractivas en expansión. Reprimarización de la economía argentina* (pp. 19–44). Buenos Aires: Antropofagia.

Toledo López, V. (2013). Los agrocombustibles como un eje del extractivismo en la Argentina. In N. Giarracca & M. Teubal (Eds.), *Actividades extractivas en expansión. ¿Reprimarización de la economía argentina?* (pp. 137–158). Buenos Aires: Antropofagia.

Toledo López, V. (2016). Agroenergía y discurso del desarrollo. Un análisis de narrativas regionales y locales a propósito de la producción de biodiesel en Santiago del Estero. In G. Merlinsky (Ed.), *Cartografías del conflicto ambiental en Argentina II* (pp. 197–226). Buenos Aires: Ediciones Ciccus.

Toledo López, V. (2017). La política agraria del kirchnerismo. Entre el espejismo de la coexistencia y el predominio del agronegocio. *Mundo Agrario*.

Toledo López, V., & Tittor, A. (2019). Contradicciones en torno a las innovaciones y certificaciones en el sector de la bioenergía en Argentina. *Letras Verdes, Revista Latinoamericana de Estudios Socioambientales, 26*, 87–110.

UNEP. (2014). Assessing Global Land Use: Balancing Consumption with Sustainable Supply. https://wedocs.unep.org/bitstream/20.500.11822/8861/1/assessing_global_land_use.pdf. Accessed 26 May 2020.

Verzeñassi, D. (2014). Agroindustria, salud y soberanía. El modelo agrosojero y su impacto en nuestras vidas. In D. Melón (Ed.), *La Patria Sojera. El modelo agrosojero en el Cono Sur* (pp. 31–48). Buenos Aires: Editorial El Colectivo.

Open Access This chapter is licensed under the terms of the Creative Commons Attribution 4.0 International License (http://creativecommons.org/licenses/by/4.0/), which permits use, sharing, adaptation, distribution and reproduction in any medium or format, as long as you give appropriate credit to the original author(s) and the source, provide a link to the Creative Commons license and indicate if changes were made.

The images or other third party material in this chapter are included in the chapter's Creative Commons license, unless indicated otherwise in a credit line to the material. If material is not included in the chapter's Creative Commons license and your intended use is not permitted by statutory regulation or exceeds the permitted use, you will need to obtain permission directly from the copyright holder.

Index

A

Agrarian extractivism 311, 313, 314, 317–320, 322, 325
Agribusiness 11, 13, 34, 38, 229, 239, 242, 247–251, 254, 270, 273, 281, 312–315
Agricultural research 28, 35, 38
Agricultural sector 13, 17, 28, 32, 36–38, 175, 279, 310, 312, 315
Agriculture 4, 5, 11, 27, 32–34, 36, 38, 39, 49, 70, 109, 117, 119, 152, 180, 181, 208, 220, 224, 226, 227, 231, 232, 239, 243, 244, 254, 278, 299, 301, 310–314, 316–322, 325, 326. *See also* Agricultural sector
Agrofuels 7, 14, 217, 239–242, 245, 246, 270–273, 281, 314, 322
Agroindustry 111, 244, 247, 251, 254, 256. *See also* Agribusiness
Alternative 10, 11, 13, 14, 17, 18, 26, 28, 38–40, 46, 59, 87, 119, 152, 154, 156, 158, 159, 222, 241, 250, 270, 287
Amazon 36–38, 158, 266, 271, 273, 279
Argentina 4, 11, 14, 16, 239, 240, 242–248, 254, 291, 296, 309–311, 315–321, 324, 325

B

Bio-Boom 299
Biodiesel 6, 14, 34, 37, 114, 157, 239, 240, 242, 244–247, 249–255
Bioeconomic imaginary 132, 133, 136–138, 143, 146

Bioeconomy agenda 7, 11, 12, 119, 121, 217, 265, 278–281
Bioeconomy conferences/symposia/summit 309, 311, 323
Bioeconomy strategy 4, 7, 12, 25–29, 32–34, 39, 65, 108–110, 112, 113, 116–118, 120–122, 131, 133, 135–138, 218, 288, 291, 300, 302
Bioelectricity 13, 114, 151–154, 156, 157, 159–166, 293, 323
Bioenergy 5, 7, 8, 10–12, 14, 27, 34, 57, 85, 86, 94, 96–99, 107–110, 112–121, 137, 138, 141, 152, 155, 156, 159, 175, 221, 269, 272, 276, 287, 293, 310, 323
Bioenergy cooperatives 120
Bioethanol 14, 114, 157, 217, 218, 221, 222, 323
Biofuels 13, 34, 49, 53–57, 97, 98, 118, 151, 155, 230, 239, 270, 271, 272, 293, 300, 310, 320. *See also* Agrofuels
Biofuture 164, 270–273
Biomass 4, 5, 7, 8, 10–12, 14, 15, 25, 27, 28, 37, 52, 56, 57, 92, 97, 98, 107–109, 114–118, 137–139, 152, 155, 159–163, 196, 197, 232, 240, 266, 267, 269–274, 280, 287–293, 295–303, 310, 317, 321, 322
Biomass flows 295–299, 302
Bio-Modesty 299, 301
Bio-patents 79, 98
Bio-Scarcity 299, 301
Biotechnology 4, 5, 11, 16, 25, 31, 34, 50, 69–71, 116, 119, 269, 270, 276, 312, 317, 318, 325, 326
Brazil 11, 13, 15, 26, 31, 32, 34–39, 151, 152, 154–159, 162, 217–219, 221–223, 227, 265–271, 273, 276–281, 290, 291, 296
BRICS 15, 267, 268, 290

C

Canada 54, 56, 58, 267, 292
Carbon emissions 98, 138, 271, 274
Carbon neutrality/carbon sinks/carbon storage 132–134, 137–146, 266, 271, 273. *See also* Carbon emissions
Cheap labour 182, 200–202
China 15, 30, 31, 34, 36, 144, 179, 265–268, 274–281, 290, 291
Citizenship 6, 14, 199–201, 209
Civil society 12, 18, 39, 108, 118, 155, 218, 271
Class 6, 14, 17, 31, 48, 96, 98, 198, 241, 278
Climate change 3, 9, 12, 17, 25, 27, 45–47, 49, 111, 112, 131, 133, 136, 143, 144, 155, 156, 160, 269, 270, 302, 317
Climate mitigation. *See* Climate change
Co-construction of markets and nature 46, 52, 57, 58
Colonial(ism) 5–7, 14, 26, 30, 32, 37, 71, 91, 92, 179–181, 187, 200, 241, 278, 311
Coloniality of gender 178, 181
Coloniality of power 177
"Coolies" 179–181

Index 333

Corn 34, 36–38, 107, 118, 248, 249, 315, 323
Cropland 297, 298, 301

D

Degradation 6, 14, 85, 93, 196, 250, 254, 301, 302, 312, 313, 319
Disciplined labour 176–178, 186–188
Division of labour 30, 33, 178, 201, 208, 289–291

E

Ecological civilization 266, 274–276, 280, 281
Ecological distribution 241, 249, 251, 253, 255
Ecological modernization 9, 255, 323
Efficiency 28, 38, 77, 92, 94, 95, 115, 118, 159–161, 164, 303
Energy 5, 7, 9, 10, 13, 18, 28, 52, 65, 85–97, 99, 107–112, 114–122, 137, 138, 151–165, 186, 196, 240, 246, 265, 267–275, 280, 287, 292, 293, 297, 299, 303, 314, 322, 323
Energy cooperatives 115, 119
Energy demand 115, 117, 267, 268
Energy epistemics 109, 111, 113, 121
Energy justice 12, 18, 108–112, 119–121
Energy security 155, 158, 159, 162, 165, 269
Entropy 88, 94

Environmental appropriation 241, 249, 251, 253
Ethnicity 6, 198
Eurocentrism/eurocentric 8, 29
European plant patent 67, 77
European seed production 79
European Union (EU) 4, 9, 11, 15, 16, 26, 28, 31–33, 52, 55, 68, 116, 133, 134, 138, 139, 142–145, 154, 245, 287–293, 295–303, 317
Exclusion(s) 14, 38, 69, 109–113, 118, 120, 121, 222, 231, 247
Exploitation 17, 85, 92, 178, 181, 188, 196–199, 210, 220, 314
Extraction 7, 10, 15, 16, 26, 30, 35, 38, 79, 88, 91, 96, 199, 266, 279–281, 291, 311, 313, 314, 316, 319
Extractive knowledge 9, 26, 32, 36, 38, 39
Extractive relations 16, 39, 292, 297, 298, 301–303
Extractivism 9, 17, 32, 240, 289, 291, 292, 303, 310–315

F

"Falls" 70, 86, 90, 92, 93
Feminist theory 176, 185
Finland 11, 13, 132–139, 141–146
First generation, second generation 7
Fodder 5, 114, 118, 293, 295, 296, 310, 314
Food-or-fuel-debate 118
Forest 13, 17, 50, 51, 56, 58, 59, 91, 94, 118, 132–135, 137–146, 248, 249, 254, 281, 301, 315
Forest policy regime 132–135, 146

Forest sector/forest industry/forestry 4–7, 11, 13, 49, 115, 117, 132–136, 138–146, 248, 298
Fossil 3, 13, 97–99, 112, 116, 131, 152, 156, 196, 265, 266, 270, 273, 274, 276, 280, 281, 287, 293, 298, 313
Framing and reframing 11, 38, 46, 131–133, 154, 276, 317, 325

G
Gender 6, 11, 96, 176, 178, 187, 188, 199
Gendered division of labour 184, 203
Genetically modified organisms (GMOs) 10, 243, 315, 317, 318, 324, 325
Germany 4, 8, 9, 11, 12, 26, 32, 33, 36, 55, 80, 107, 110, 113–115, 117, 118, 121, 267
Global energy consumption 268, 272
Global inequalities 7, 10, 15, 26, 30–32, 36, 99, 265, 281
Global knowledge production 15, 29, 31, 33
Glyphosate 243, 315, 316. *See also* Pesticides

H
Household income 204, 206
Housewifisation 178, 180
Hydroelectricity 93, 158, 160

I
Inclusion(s) 12, 109–113, 119, 121, 161, 248, 321, 325
Indigenous people 6, 14, 38, 93, 96, 187, 223, 226–228, 230, 254
Indonesia 11, 14, 98, 175–179, 182–184, 187, 188, 196, 202, 296, 298
Industrial development 155, 165
Industrialization 157, 246, 250, 279, 313, 322
Innovation(s) 10, 16, 18, 27, 66, 71, 74–76, 79, 110, 111, 113, 117, 121, 135, 160, 164, 218, 310, 317
Intersectionality 9
Investment 10, 66, 69, 70, 79, 87, 109, 114, 120, 136, 137, 142, 153, 159–161, 163–165, 221, 222, 228, 250, 270, 275, 280, 325
Iron ore 277–279, 281

K
Knowledge 6, 7, 12, 17, 18, 25–27, 29–35, 38, 66, 67, 89, 109, 111, 112, 119–122, 153, 208, 275, 319, 323
Knowledge production 5, 9, 12, 26, 28–35, 37–40, 110, 111, 113, 121

L
Labour 6, 10, 13–15, 17, 66, 87, 92, 94, 97, 109, 176–184, 186–188, 196–201, 204–206, 208–210, 218–221, 223, 224,

226, 227, 230–232, 254, 313, 320
Labour migration 17, 197, 199–202, 208–210
Labour regime(s) 176, 177, 181, 182, 208, 219–221, 223, 231
Land access 28, 37, 221, 231
Land dispossession 178, 187
Land footprint 272, 298
Land use (change) 28, 138, 139, 196, 240, 243, 246, 279, 297, 301, 303, 312, 314, 316
Little-e/little energies 18, 91, 95, 97
Lobbying 142, 146, 154
LULUCF 138, 139, 141, 142, 144. *See also* Land use (change)

M

Malaysia 4, 11, 14, 98, 182, 196, 197, 200–209, 291, 296
Market development policies (MDPs) 53–56
Market(s) 10, 14, 16, 25, 30, 34–36, 46–54, 56–58, 67, 69, 70, 74–76, 78, 79, 116, 152, 153, 155–160, 162–165, 196, 200, 201, 203, 207, 221, 239, 244–247, 271, 278, 281, 290–292, 295, 312, 314, 323, 325
Marxism 48, 177
Material Flow Accounts (MFA) 289
Material flows 291–293, 298, 301, 303, 316
Materialities 10, 50–52, 54, 56–58, 112
Millennium Development Goals (MDGs) 154

Monocropping/monocultures 107, 114, 115, 248, 279, 281, 292, 310–312, 314, 315, 318–320, 324–326

N

Native traits 10, 68–70, 72, 73, 75, 76, 78, 79
Natural gas 151, 152, 158, 160–162, 165, 270
Neoliberal bioeconomy(ies) 47, 52
Neoliberalism/neoliberalization 46–51, 57
Neoliberal natures 9, 46, 49–51, 57, 58
Nyai 180, 181

O

Oil 13, 37, 90, 91, 95–98, 156–158, 175, 176, 178, 179, 182–184, 186–188, 196, 198, 200, 202, 203, 244, 246, 252, 267, 269, 270, 272, 273, 276–278, 280, 281, 291, 296, 298, 300, 322

P

Palm oil 6, 14, 175, 176, 188, 196, 197, 199, 202, 203, 206, 208–210, 295, 296, 298, 302
Paris Climate Agreement 132. *See also* Climate change
Patent economy 77
Patented traits 71, 72, 74, 77
Patents 10, 28, 31, 36, 66–79, 315
Peasants 14, 17, 18, 93, 96, 98, 180, 181, 218, 220, 223, 227–233,

248–250, 254, 312, 314, 321, 325
Pesticides 6, 36, 111, 242, 243, 248, 279, 310, 312, 315–318, 324–326
Plant patent 10, 67, 78
Political ecology (PE) 5, 6, 108, 240, 241, 255
Pré-Sal 158, 280
Primitive accumulation 177, 178, 187
Public debate 13, 18, 29, 133, 134, 139, 143, 145

R

Racial capitalism 176, 177
Refeudalisation 79
Regime 28, 39, 118, 132–146, 153, 154, 160, 163, 165, 176–178, 197, 200–202, 208, 210, 246, 250, 281
Regulations 48, 98, 113, 118, 120, 133, 134, 138, 142, 144, 145, 155, 157, 159, 163, 165, 181, 185, 186, 201, 205, 210, 247, 273
Remittances 206–208
Renewable(s) 6, 13, 25, 46, 54, 107, 110, 112–116, 118, 119, 121, 137, 138, 151, 152, 154–156, 158, 159, 161, 162, 164, 165, 196, 232, 246, 271–273, 275, 287
Reprimarization 255
Reproduction 6, 98, 153, 188, 198, 202, 205–208, 210
Reproduction of labour power 199

Research and development (R&D) 5, 12, 26, 27, 38, 54, 66, 70, 74–76, 276
Research and innovation (R&I) 4, 108–110, 113, 119–121
Resource 4–6, 8, 15–17, 25, 27, 30, 32, 35, 46, 50–52, 57, 58, 75, 88, 95, 108, 111, 115–117, 121, 122, 132, 135, 136, 141, 152, 153, 157, 160, 164, 219, 232, 241, 255, 266, 269, 273, 278, 279, 281, 287–289, 291, 293, 298, 299, 302, 303, 310, 312–314, 317
Rural labour 220

S

Scale 7, 8, 26, 30, 54, 74, 87, 108, 115, 153, 161, 217, 221, 272, 276, 279, 280, 288, 295, 303, 321
Segmented labour market 200
Semi-proletarianisation 221, 231
Social class 241
Social devaluation 198, 199, 209
Social movements 10, 32, 38, 40, 86, 95, 111, 114, 318
Social reproduction 179, 180, 210, 220
Socio-ecological inequalities 5–9, 13, 15–17, 26, 30, 37, 39
Socio-energy systems 109–112, 121
Socio-technical systems 109, 157
Soil 36, 91, 92, 254, 279, 301, 312–314, 316, 319, 325
Solar 13, 87, 117, 152, 155–162, 165, 275

South-South 15, 36, 265, 267, 268, 276, 278, 280, 281
Soy(a)/soybean 11, 16, 27, 32, 34, 36–39, 70, 224, 228, 229, 240, 242–244, 246, 248–252, 254, 271, 277–279, 281, 296, 310, 311, 313–324
Speculation 10, 79, 98
State 7, 11, 13, 17, 30, 47, 49, 50, 56, 92, 116, 118–120, 134, 136, 138, 139, 142, 155, 157, 162, 165, 177, 187, 195, 197, 200–203, 210, 218, 219, 221–223, 227, 252, 253, 255, 270, 289–291, 319
Subsistence perspective 94
Subsistence work 198, 207, 219, 220, 228, 323
Sugarcane 13, 14, 27, 34, 37, 152, 218, 221–224, 226–231, 269–273, 279
Superexploitation 197–199, 202, 203, 205, 208–210
Supply 54, 57, 91, 151, 155, 157, 161, 200, 201, 222, 245, 246, 280, 297, 299–302
Sustainability 9, 12, 16, 34, 52, 54, 55, 109, 110, 114, 132, 134, 138, 143, 145, 154–156, 158, 160, 162, 165, 248, 274, 288, 289, 303, 310, 313, 324, 325
Sustainable Development Goals (SDGs) 154, 155

T
Technological innovation 4, 25, 27, 28, 34, 37, 38, 66, 117, 319

Technologies 15, 16, 26–30, 32–35, 39, 48, 50, 53, 57, 58, 66, 67, 72–74, 108, 114, 117, 120, 158, 164, 218, 247, 272, 277, 299, 312, 314
Territorial changes 240, 241, 247, 250, 251
Territory 91, 241, 248, 254
Theory of Access 219, 231
Thermodynamic energy/laws 10, 85, 88, 89, 91, 93–99
Trade 17, 135, 243, 265, 266, 273, 276–281, 288–290, 293, 295, 298
Trait breeding 69, 78, 79
Traits 68–71, 73–75, 78, 79
Transformation 8, 18, 46, 48, 51, 68, 89, 90, 109, 115, 118, 134, 137, 144, 163, 195, 197, 242, 249, 250, 310, 311
Transition 4, 7, 12, 25, 45, 46, 53, 86, 87, 94, 99, 107–111, 113, 115, 117, 120, 121, 131, 135, 152–158, 161, 164, 165, 175, 188, 197, 232, 265, 266, 270, 273, 280, 281
Transnational corporations 315

U
Union (organizing) 209

V
Valuation of nature 241, 254, 255
Value 5, 9, 10, 46, 50, 57, 65–67, 72, 73, 75, 76, 78, 79, 87, 88, 94, 95, 112, 116, 117, 119, 154, 198, 199, 229, 232, 247,

249, 255, 277, 287, 291, 292, 313, 315, 316, 320
Value-adding 66, 141, 244, 250, 277, 287, 320

W

Wage labour 14, 95, 177, 198, 199, 207, 220, 221, 227, 231
Waste 37, 88, 91–94, 114, 115, 117, 251, 287, 322–324
Water pollution 252, 254
Wind 13, 88, 91, 95, 133, 146, 151, 152, 155–162, 165, 275
Wood 86, 99, 107, 114, 120, 134, 138, 141, 293, 296, 300
World systems theory (WST) 6, 17, 26, 29–31, 289, 290